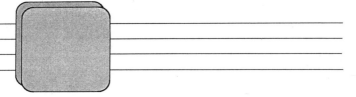

MW00354764

STUDY GUIDE
VOLUME TWO
FOR STEWART'S
CALCULUS

THIRD EDITION

RICHARD ST. ANDRE

Central Michigan University

BROOKS/COLE PUBLISHING COMPANY

I(T)P An International Thomson Publishing Company

Pacific Grove • Albany • Bonn • Boston • Cincinnati • Detroit • London • Madrid • Melbourne
Mexico City • New York • Paris • San Francisco • Singapore • Tokyo • Toronto • Washington

Sponsoring Editor: Elizabeth Rammel
Editorial Assistant: Carol Benedict
Production Coordinator: Dorothy Bell
Cover Design: Vernon T. Boes
Cover Sculpture: Christian Haase
Cover Photo: Ed Young
Typesetting: Laurel Technical Services
Printing and Binding: Malloy Lithographing, Inc.

For more information, contact:

BROOKS/COLE PUBLISHING COMPANY
511 Forest Lodge Rd.
Pacific Grove, CA 93950
USA

International Thomson Publishing Europe
Berkshire House 168-173
High Holborn
London WC1V 7AA
England

Thomas Nelson Australia
102 Dodds Street
South Melbourne, 3205
Victoria, Australia

Nelson Canada
1120 Birchmount Road
Scarborough, Ontario
Canada M1K 5G4

International Thomson Editores
Campos Eliseos 385, Piso 7
Col. Polanco
11560 México D. F. México

International Thomson Publishing GmbH
Königwinterer Strasse 418
53227 Bonn
Germany

International Thomson Publishing Asia
221 Henderson Road
#05–10 Henderson Building
Singapore 0315

International Thomson Publishing Japan
Hirakawacho Kyowa Building, 3F
2-2-1 Hirakawacho
Chiyoda-ku, Tokyo 102
Japan

Printed in the United States of America

10 9 8 7 6 5 4 3

ISBN 0-534-21805-9

Preface

This Study Guide is designed to supplement Chapters 10 – 15 of Calculus, 3rd edition, by James Stewart. It may also be used with Multivariable Calculus, 3rd edition.

Your text is well written in a very complete and patient style. This Study Guide is not intended to replace it. You should read the relevant sections of the text and work problems, lots of problems. Calculus is not a spectator sport.

This Study Guide captures the main points and formulas of each section and provides short, pithy questions that will help you understand the essential concepts. Every question has an explained answer. The pages are perforated so that you can detach them. The two column format allows you to cover the answer portion of a question while working on it and then uncover the given answer to check your solution. Working in this fashion leads to greater success than simply perusing the solutions.

As a quick check of your understanding of a section work the page of On Your Own questions located toward the back of the Study Guide. These are all multiple choice questions -- the type that you might see on an exam in a large-sized calculus class. You are "on your own" in the sense that an answer, but no solution, is provided for each question.

I hope that you find this Study Guide helpful in understanding the concepts and solving the exercises in Calculus, 3rd edition.

Richard St. Andre

Contents

Contents (Continued)

Contents (Continued)

Please cut page down center line.
Use the half page to cover the right
column while you work in the left.

10

Infinite Sequences and Series

"I'M BEGINNING TO UNDERSTAND ETERNITY, BUT INFINITY IS STILL BEYOND ME."

Cartoons courtesy of Sidney Harris. Used by permission.

Sequences

Concepts to Master

A. Sequ̶e̶ sequences (increasing, decreasing); Bounded sequences; Sequences
, Limit of a sequence; Convergent; Divergent
B. M recursively

Summary and Focus Questions

A <u>sequence</u> is a list of numbers arranged in order:

$$\{a_n\} = a_1,\ a_2,\ a_3,\ \ldots,\ a_n,\ a_{n+1},\ \ldots$$

a_n is the <u>nth term</u> of the sequence.

$\lim\limits_{n\to\infty} a_n = L$ means the sequence $\{a_n\}$ <u>converges to L</u>, that is, as n grows larger the a_n values get closer and closer to L. Formally, $\lim\limits_{n\to\infty} a_n = L$ means for all $\varepsilon > 0$ there exists a positive integer N such that $\left|a_n - L\right| < \varepsilon$ for all $n > N$. If the limit does not exist, $\{a_n\}$ <u>diverges</u>.

$\lim\limits_{n\to\infty} a_n = \infty$ means the a_n terms grow without bound.

One way to evaluate $\lim\limits_{n\to\infty} a_n$ is to find a real function $f(x)$ such that $f(n) = a_n$ for all n. If $\lim\limits_{n\to\infty} f(x) = L$, then $\lim\limits_{n\to\infty} a_n = L$. The converse is false.

All limit laws for limits of functions at infinity are valid for convergent sequences. A version of the Squeeze Theorem also holds:

If $a_n \le b_n \le c_n$ for all $n \ge N$ and $\lim\limits_{n\to\infty} a_n = \lim\limits_{n\to\infty} c_n = L$, then $\lim\limits_{n\to\infty} b_n = L$.

1) Find the fourth term of the sequence $a_n = \frac{(-1)^n}{n^2}$.

$a_4 = \frac{(-1)^4}{4^2} = \frac{1}{16}$.

2) Determine whether each converges.

a) $a_n = \frac{n}{n^2+1}$.

$\lim\limits_{n\to\infty} \frac{n}{n^2+1} = $ (divide by n^2)

$\lim\limits_{n\to\infty} \frac{\frac{1}{n}}{1+\frac{1}{n^2}} = \frac{0}{1+0} = 0.$

Thus a_n converges to 0.

b) $b_n = \frac{n^2+1}{2n}$.

b_n grows without bound so $\{b_n\}$ diverges.
We may write $\lim\limits_{n\to\infty} \frac{n^2+1}{2n} = \infty.$

c) $c_n = \frac{(-1)^n n}{n+1}$.

The sequence $\frac{n}{n+1}$ converges to 1 so $c_n = \frac{(-1)^n n}{n+1}$ is alternating values near 1 and -1. Thus $\{c_n\}$ diverges.

3) Find $\lim\limits_{n\to\infty} \frac{\cos n}{n}$.

0. Since $-1 \le \cos n \le 1$, $-\frac{1}{n} \le \frac{\cos n}{n} \le \frac{1}{n}$.
Thus as $n \to \infty$, $\frac{\cos n}{n} \to 0.$

4) Find $\lim\limits_{n\to\infty} \frac{\ln n}{n}$.

Let $f(x) = \frac{\ln x}{x}$.

$\lim\limits_{x\to\infty} \frac{\ln x}{x} \left(= \frac{0}{0} \text{ form} \right)$

$= \lim\limits_{x\to\infty} \frac{1/x}{1}$ (L'Hôpital's Rule) $= 0.$

Thus $\lim\limits_{x\to\infty} \frac{\ln n}{n} = 0.$

B. $\{a_n\}$ is <u>increasing</u> means $a_{n+1} \ge a_n$ for all n.
$\{a_n\}$ is <u>decreasing</u> means $a_{n+1} \le a_n$ for all n.
$\{a_n\}$ is <u>monotonic</u> if it is either increasing or decreasing.
One way to show $\{a_n\}$ is increasing is to show that $\frac{da_n}{dn} \ge 0$ (treating n as a real number).
$\{a_n\}$ is <u>bounded above</u> means $a_n \le M$ for some M and all n.
$\{a_n\}$ is <u>bounded below</u> means $a_n \ge m$ for some m and all n.
$\{a_n\}$ is <u>bounded</u> if it is both bounded above and bounded below.

An important way to show a sequence converges is:
  If $\{a_n\}$ is bounded and monotonic, then $\{a_n\}$ converges. The converse is
  false.
Also:
  If $\{a_n\}$ converges, then $\{a_n\}$ is bounded. The converse is false.

A sequence $\{a_n\}$ is defined by a recurrence relation means that a_1 is defined
and then a_{n+1} is defined in terms of previous values, for $n = 1,\ 2,\ 3\ldots$

For example, $\{a_n\}$ given by
$$a_1 = \tfrac{1}{2}$$
$$a_{n+1} = \tfrac{a_n}{2},\ n = 1,\ 2,\ 3\ldots$$
is the sequence $\tfrac{1}{2},\ \tfrac{1}{4},\ \tfrac{1}{8},\ \tfrac{1}{16},\ \ldots$. This is another way that $a_n = \tfrac{1}{2^n}$ may be
defined.

5) Is $c_n = \tfrac{1}{3n}$ bounded above? bounded
   below?

$\left\{\tfrac{1}{3n}\right\}$ is bounded above by $\tfrac{1}{3}$ and bounded
below by 0.

6) Is $s_n = \tfrac{n}{n+1}$ an increasing sequence?

Yes, because $\tfrac{ds_n}{dn} = \tfrac{(n+1)-n}{(n+1)^2} = \tfrac{1}{(n+1)^2} > 0.$

7) True or False:

   a) If $\{a_n\}$ is not bounded below then
      $\{a_n\}$ diverges.

   True.

   b) If $\{a_n\}$ is decreasing and $a_n \geq 0$
      for all n then $\lim\limits_{n\to\infty} a_n$ exists.

   True.

   c) If $\{a_n\}$ and $\{b_n\}$ converge, then
      $\{a_n + b_n\}$ converges.

   True.

   d) If $\{a_n\}$ and $\{b_n\}$ diverge, then
      $\{a_n + b_n\}$ diverges.

   False. For example, $a_n = n^2$ and $b_n = -n^2$
   diverge.

8) If $\lim\limits_{n\to\infty} s_n = 4$ and $\lim\limits_{n\to\infty} t_n = 2$, then

   a) $\lim\limits_{n\to\infty} (8s_n - 2t_n) = \underline{\qquad}.$

   $8(4) - 2(2) = 28.$

b) $\lim\limits_{n\to\infty} \frac{3s_n}{t_n} =$ _____.

9) Define the sequence
    $\frac{1}{2}, \frac{3}{4}, \frac{7}{8}, \frac{15}{16}, \frac{31}{32}, \dots$ with a recurrence relation.

10) Suppose $\{a_n\}$ is defined as $a_1 = 2$, $a_2 = 1$ and $a_{n+2} = a_{n+1} - a_n$ for $n = 1, 2, 3, \dots$. Find $\lim\limits_{n\to\infty} a_n$.

$\frac{3(4)}{2} = 6.$

$a_1 = \frac{1}{2}; a_{n+1} = a_n + \frac{1}{2^{n+1}}$, for $n = 1, 2, 3,$ $\dots$ $\left(\text{Note: a non-recursive answer is}\right.$
$a_n = \frac{2^n - 1}{2^n}.\Big)$

Calculate a few terms to understand the pattern.
$a_1 = 2, a_2 = 1$
$a_3 = 1 - 2 = -1$
$a_4 = -1 - 1 = -2$
$a_5 = -2 - (-1) = -1$
$a_6 = -1 - (-2) = 1$
$a_7 = 1 - (-1) = 2$
$a_8 = 2 - 1 = 1$
$a_9 = 1 - 2 = -1$
$\vdots$

The terms cycle through $1, -1, \dots$ and eventually return to $1, -1$. $\lim\limits_{n\to\infty} a_n$ does not exist.

Series

Concepts to Master

A. Infinite series; Partial sums; Convergent and divergent series; Convergent series laws

B. Geometric series; Value of a converging geometric series; Harmonic series

Summary and Focus Questions

A. Adding up all the terms of a sequence $\{a_n\}$ is an <u>(infinite) series</u>:

$$\sum_{k=1}^{\infty} a_k = a_1 + a_2 + a_3 + \ldots + a_n + \ldots$$

If we stop adding after n terms, we have the <u>nth partial sum</u> of the series:

$$s_n = \sum_{k=1}^{\infty} a_k = a_1 + a_2 + a_3 + \ldots + a_n + \ldots$$

$\sum_{n=1}^{\infty} a_n$ <u>converges (to s)</u> means $\lim\limits_{n \to \infty} s_n$ exists (and is s). Thus a series converges if the limit of its sequence of partial sums exists.

If that limit does not exist, then $\sum_{n=1}^{\infty} a_n$ <u>diverges</u>.

If $\sum_{n=1}^{\infty} a_n$ converges, then $\lim\limits_{n \to \infty} a_n = 0$. Rewritten in a contrapositive form, this is the <u>Test for Divergence</u>:

If $\lim\limits_{n \to \infty} a_n \neq 0$, $\sum_{k=1}^{\infty} a_k$ diverges.

The converse is false - just because the terms a_n get small is not enough to conclude the series converges.

If $\displaystyle\sum_{n=1}^{\infty} a_n$ converges (to L) and $\displaystyle\sum_{n=1}^{\infty} b_n$ converges (to M) then:

$$\sum_{n=1}^{\infty} (a_n \pm b_n) \text{ converges (to } L \pm M\text{)}.$$

$$\sum_{n=1}^{\infty} ca_n \text{ converges (to } cL\text{) for } c \text{ any constant.}$$

1) A series converges if the _____ of the sequence of _____ exists.

limit, partial sums

2) Find the first four partial sums of $\displaystyle\sum_{k=1}^{\infty} \frac{1}{k^2}$.

$s_1 = \frac{1}{1^2} = 1$.

$s_2 = \frac{1}{1^2} + \frac{1}{2^2} = 1 + \frac{1}{4} = \frac{5}{4}$.

$s_3 = \frac{1}{1^2} + \frac{1}{2^2} + \frac{1}{3^2} = \frac{5}{4} + \frac{1}{9} = \frac{49}{36}$.

$s_4 = \frac{1}{1^2} + \frac{1}{2^2} + \frac{1}{3^2} + \frac{1}{4^2} = \frac{49}{36} + \frac{1}{16} = \frac{205}{144}$.

3) Sometimes, Always, or Never:
If $\displaystyle\lim_{n\to\infty} a_n = 0$, $\displaystyle\sum_{n=1}^{\infty} a_n$ converges.

Sometimes.

4) Suppose $\displaystyle\sum_{n=1}^{\infty} a_n = 3$ and $\displaystyle\sum_{n=1}^{\infty} b_n = 4$.
Evaluate each of the following:

a) $\displaystyle\sum_{n=1}^{\infty} (a_n + 2b_n)$

$3 + 2(4) = 11$.

b) $\displaystyle\sum_{n=1}^{\infty} \frac{a_n}{5}$

$\frac{3}{5}$.

c) $\displaystyle\sum_{n=1}^{\infty} \frac{1}{a_n}$

This diverges because if $\displaystyle\sum_{n=1}^{\infty} a_n = 3$, then $\displaystyle\lim_{n\to\infty} a_n = 0$. Thus $\displaystyle\lim_{n\to\infty} \frac{1}{a_n} \neq 0$, so $\displaystyle\sum_{n=1}^{\infty} \frac{1}{a_n}$ diverges.

5) Sometimes, Always, or Never:

a) $\sum\limits_{n=1}^{\infty} a_n$ converges, but $\lim\limits_{n\to\infty} a_n$ does

not exist.

Never. If $\sum\limits_{n=1}^{\infty} a_n$ converges, then $\lim\limits_{n\to\infty} a_n$ exists and is 0.

b) $\sum\limits_{n=1}^{\infty} a_n$ converges, $\sum\limits_{n=1}^{\infty} (a_n + b_n)$

converges, but $\sum\limits_{n=1}^{\infty} b_n$ diverges.

Never. $\sum\limits_{n=1}^{\infty} b_n = \sum\limits_{n=1}^{\infty} (a_n + b_n) - \sum\limits_{n=1}^{\infty} a_n$ is the difference of two convergent series and must converge.

6) Does $\sum\limits_{n=1}^{\infty} \cos\left(\frac{1}{n}\right)$ converge?

No, the nth term does not approach 0.

B. A <u>geometric series</u> has the form
$$\sum_{n=1}^{\infty} ar^{n-1} = a + ar + ar^2 + ar^3 + \dots$$
and converges (for $a \neq 0$) to $\frac{a}{1-r}$ if and only if $-1 < r < 1$.

The <u>harmonic series</u> $\sum\limits_{n=1}^{\infty} \frac{1}{n} = 1 + \frac{1}{2} + \frac{1}{3} + \dots + \frac{1}{n} + \dots$ diverges.

7) a) $\sum\limits_{n=1}^{\infty} \frac{2^n}{100}$ is a geometric series in
which $a =$ _____ and $r =$ _____.

$\sum\limits_{n=1}^{\infty} \frac{2^n}{100} = \frac{2}{100} + \frac{4}{100} + \frac{8}{100} + \dots$
so $a = \frac{2}{100}$ and $r = 2$.

b) Does it converge?

No, since $r \geq 1$.

8) $\sum\limits_{n=1}^{\infty} \left(\frac{-2}{9}\right)^n =$ _____.

A geometric series with $a = r = -\frac{2}{9}$ which
converges to $\frac{a}{1-r} = -\frac{2}{11}$.

9) Does $\sum\limits_{n=1}^{\infty} \frac{6}{n}$ converge?

It diverges, since it is a multiple of the harmonic series.

10) $\displaystyle\sum_{n=1}^{\infty} \frac{2^n}{3^{n+1}} =$ _____.

$\displaystyle\sum_{n=1}^{\infty} \frac{2^n}{3^{n+1}} = \sum_{n=1}^{\infty} \frac{2}{9}\left(\frac{2}{3}\right)^{n-1}$, which is a geometric series with $a = \frac{2}{9}$, $r = \frac{2}{3}$ that converges to $\dfrac{\frac{2}{9}}{1 - \frac{2}{3}} = \frac{2}{3}$.

11) Achilles gives the tortoise a 100 m head start. If Achilles runs at 5 m/sec and the tortoise at $\frac{1}{2}$ m/sec how far has the tortoise traveled by the time Achilles catches him?

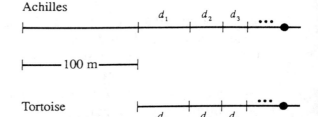

Let $d_1 =$ distance tortoise traveled while Achilles was running to the tortoise's starting point.

$d_1 = \left(\frac{100\,\text{m}}{5\,\text{m/sec}}\right)\left(\frac{1}{2}\,\text{m/sec}\right) = 10$ m.

Let $d_2 =$ distance tortoise traveled while Achilles was running the distance d_1. Since $d_1 = 10$ and Achilles runs at 5 m/sec $d_2 = \left(\frac{10\,\text{m}}{5\,\text{m/sec}}\right)\frac{1}{2}$ m/sec $= 1$ m.

For each n, $d_n = \left(\frac{d_{n-1}}{5}\right)\frac{1}{2} = \frac{d_{n-1}}{10}$.

The total distance traveled by the tortoise is $\displaystyle\sum_{n=1}^{\infty} d_n = 10 + 1 + \frac{1}{10} + \ldots$ which is a geometric series with $a = 10$, $r = \frac{1}{10}$. This converges to $\dfrac{10}{1 - \frac{1}{10}} = \frac{100}{9}$ m.

The Integral Test

Concepts to Master

A. Integral Test for convergence; p-series
B. Estimate the sum of a convergent series

Summary and Focus Questions

This section and sections 10.4, 10.5, and 10.6 provide tests to determine whether a series converges. The tests will not give a value for the series, only whether there is such a value, and in a few cases a good estimate if the value exists.

A. <u>Integral Test</u>: Let $\{a_n\}$ be a sequence. Suppose there is $f(x)$, a positive, continuous, and decreasing function on $[1, \infty)$ such that $a_n = f(n)$ for all n. Then $\sum\limits_{n=1}^{\infty} a_n$ converges if and only if $\int_1^{\infty} f(x)dx$ converges.

The Integral Test works when a_n has the form of a function whose antiderivative is easily found. It is one of the few general tests that gives necessary and sufficient conditions for convergence.

A <u>p-series</u> has the form $\sum\limits_{n=1}^{\infty} \frac{1}{n^p}$, ($p$ a constant), and converges (by the Integral Test) if and only if $p > 1$.

1) True or False:
    The Integral Test will determine to what value a series converges.

    False. The test only indicates whether a series converges.

2) Test $\sum\limits_{n=1}^{\infty} \frac{n}{e^n}$ for convergence.

$\frac{n}{e^n}$ suggests the function $f(x) = xe^{-x}$ when on $[1, \infty)$ is continuous and decreasing.
$\int_1^\infty xe^{-x} = \lim\limits_{t\to\infty} \int_1^t xe^{-x}\, dx.$

Using integration by parts,
$$\left(\begin{array}{cc} u = x & dv = e^{-x}\, dx \\ du = dx & \,v = -e^{-x} \end{array} \right)$$
$\int_1^t xe^{-x}\, dx = -xe^{-x} - e^{-x} \Big|_1^t = \frac{2}{e} - \frac{t+1}{e^t}.$
$\lim\limits_{t\to\infty} \left(\frac{2}{e} - \frac{t+1}{e^t} \right) = \frac{2}{e} - 0 = \frac{2}{e}.$
$\left(\text{By L'Hôpital's Rule, } \frac{t+1}{e^t} \to 0. \right)$
Thus $\int_1^\infty xe^{-x}\, dx = \frac{2}{e}$, so $\sum\limits_{n=1}^{\infty} \frac{n}{e^n}$ converges.

Note: We can<u>not</u> conclude than $\sum\limits_{n=1}^{\infty} \frac{n}{e^n} = \frac{2}{e}$.

3) Which of these converge?

a) $\sum\limits_{n=1}^{\infty} \frac{1}{n^2}.$

Converges (p-series with $p = 2$).

b) $\sum\limits_{n=1}^{\infty} \frac{1}{\sqrt{n}}.$

Diverges $\left(p\text{-series with } p = \frac{1}{2} \right).$

c) $\sum\limits_{n=1}^{\infty} \frac{3}{2n^3}.$

Converges. $\sum\limits_{n=1}^{\infty} \frac{3}{2n^3} = \frac{3}{2}\sum\limits_{n=1}^{\infty} \frac{1}{n^3}$ is a multiple of a p-series with $p = 3$.

d) $\sum\limits_{n=1}^{\infty} \left(\frac{1}{n^3} + \frac{1}{8^n} \right).$

Converges. $\sum\limits_{n=1}^{\infty} \left(\frac{1}{n^3} + \frac{1}{8^n} \right) = \sum\limits_{n=1}^{\infty} \frac{1}{n^3} + \sum\limits_{n=1}^{\infty} \frac{1}{8^n}$, the sum of a convergent p-series $(p = 3)$ and a convergent geometric series $\left(r = \frac{1}{8} \right).$

B. If $\sum a_n = s$ is determined to converge by the Interal Test using $f(x)$ then the error $R_n = s - s_n$ between the series value and the nth partial sum satisfies
$$\int_{n+1}^\infty f(x)dx \le s - s_n \le \int_n^\infty f(x)dx.$$

4) a) Find the sixth partial sum of
$$\sum\limits_{n=1}^{\infty} \frac{1}{n^2}.$$

$s_6 = 1 + \frac{1}{4} + \frac{1}{9} + \frac{1}{16} + \frac{1}{25} + \frac{1}{36}$
$= \frac{5369}{3600} \approx 1.491.$

b) Estimate the difference between your answer to part a) and the exact value of $s = \sum\limits_{n=1}^{\infty} \frac{1}{n^2}$.

We know by the Integral Test using $f(x) = \frac{1}{x^2}$ that $\sum\limits_{n=1}^{\infty} \frac{1}{n^2}$ converges. Thus

$\int_7^\infty \frac{1}{x^2}\, dx \leq s - s_6 \leq \int_6^\infty \frac{1}{x^2}\, dx$.

$\int_7^\infty \frac{1}{x^2}\, dx = \lim\limits_{t\to\infty} \int_7^t x^{-2}\, dx$

$\qquad = \lim\limits_{t\to\infty} \left(\frac{1}{7} - \frac{1}{t} \right) = \frac{1}{7}$.

Likewise $\int_6^\infty \frac{1}{x^2}\, dx = \frac{1}{6}$.

Therefore $\frac{1}{7} \leq s - s_6 \leq \frac{1}{6}$.

c) Estimate $s = \sum\limits_{n=1}^{\infty} \frac{1}{n^2}$ using the results of parts a) and b).

$\frac{1}{7} \leq s - s_6 \leq \frac{1}{6}$.

$\frac{1}{7} + s_6 \leq s \leq \frac{1}{6} + s_6$.

$\frac{1}{7} + \frac{5369}{3600} \leq s \leq \frac{1}{6} + \frac{5369}{3600}$.

$\frac{41183}{25200} \leq s \leq \frac{5969}{3600}$

$1.634 \leq s \leq 1.658$

It turns out that $s = \frac{\pi^2}{6} \approx 1.645$, so s_6 is a rather good estimate.

The Comparison Test

Concepts to Master

A. Comparison Test; Limit Comparison Test
B. Estimate the sum of a convergent series

Summary and Focus Questions

A. The following tests may be used to determine the convergence of a series whose *terms are all positive*.

<u>Comparison Test:</u>

1. If there is a convergent series $\sum\limits_{n=1}^{\infty} b_n$ such that $0 \leq a_n \leq b_n$ for all $n \geq N$, then $\sum\limits_{n=1}^{\infty} a_n$ converges.

2. If there is a divergent series $\sum\limits_{n=1}^{\infty} b_n$ such that $0 \leq b_n \leq a_n$ for all $n \geq N$, then $\sum\limits_{n=1}^{\infty} a_n$ diverges.

<u>Limit Comparison Test:</u>

Let $\sum\limits_{n=1}^{\infty} a_n$ and $\sum\limits_{n=1}^{\infty} b_n$ be positive term series.

1. If $\lim\limits_{n\to\infty} \frac{a_n}{b_n} = c > 0$, then $\sum\limits_{n=1}^{\infty} a_n$ and $\sum\limits_{n=1}^{\infty} b_n$ either both converge or both diverge.

2. If $\lim\limits_{n\to\infty} \frac{a_n}{b_n} = 0$, and $\sum\limits_{n=1}^{\infty} b_n$ converges, then $\sum\limits_{n=1}^{\infty} a_n$ converges.

3. If $\lim\limits_{n\to\infty} \frac{a_n}{b_n} = 0$, and $\sum\limits_{n=1}^{\infty} b_n$ diverges, then $\sum\limits_{n=1}^{\infty} a_n$ diverges.

To successfully use either test on a series $\sum\limits_{n=1}^{\infty} a_n$, you must come up with another series $\sum\limits_{n=1}^{\infty} b_n$ (geometric, p-series, ...) that you know is convergent or divergent and compare a_n to b_n.

1) Determine whether or not each of the following converge. Find a series $\sum\limits_{n=1}^{\infty} b_n$ to use with the Comparison Test or Limit Comparison Test.

a) $\sum\limits_{n=1}^{\infty} \frac{1}{n^2+2n}$,

$b_n = $ _____ .

You must have a hunch beforehand whether $\sum\limits_{n=1}^{\infty} \frac{1}{n^2+2n}$ converges. Because $\frac{1}{n^2+2n}$ is "like" $\frac{1}{n^2}$ for large n, and $\sum\limits_{n=1}^{\infty} \frac{1}{n^2}$ is a converging p-series, our hunch is the given series converges. Now we have an idea of what kind of b_n to look for.

Use $b_n = \frac{1}{n^2}$. Since $\frac{1}{n^2+2n} < \frac{1}{n^2}$ and $\sum\limits_{n=1}^{\infty} \frac{1}{n^2}$ converges, $\sum\limits_{n=1}^{\infty} \frac{1}{n^2+2n}$ converges by the Comparison Test.

b) $\sum\limits_{n=1}^{\infty} \frac{\sqrt[3]{n}}{n+4}$,

$b_n = $ _____ .

Your hunch should be $\frac{\sqrt[3]{n}}{n+4}$ is "like" $\frac{\sqrt[3]{n}}{n} = \frac{1}{\sqrt[3]{n^2}}$. Use $b_n = \frac{1}{\sqrt[3]{n^2}}$. Then

$$\lim_{n\to\infty} \frac{a_n}{b_n} = \lim_{n\to\infty} \frac{\frac{\sqrt[3]{n}}{n+4}}{\frac{1}{\sqrt[3]{n^2}}} = \lim_{n\to\infty} \frac{n}{n+4} = 1.$$

Since $\sum\limits_{n=1}^{\infty} \frac{1}{\sqrt[3]{n^2}}$ diverges $\left(\text{a } p\text{-series with}\right.$ $\left. p = \frac{2}{3}\right)$, $\sum\limits_{n=1}^{\infty} \frac{\sqrt[3]{n}}{n+4}$ diverges by the Limit Comparison Test.

c) $\sum\limits_{n=1}^{\infty} \frac{1}{n+2^n}$,

   $b_n = $ _____.

Use $b_n = \frac{1}{2^n}$. Then for $n \geq 1$, $\frac{1}{n+2^n} \leq \frac{1}{2^n}$. Since $\sum\limits_{n=1}^{\infty} \frac{1}{2^n}$ converges $\left(\text{a geometric series}\right.$ with $\left. r = \frac{1}{2}\right)$, $\sum\limits_{n=1}^{\infty} \frac{1}{n+2^n}$ converges.

2) Suppose $\{a_n\}$ and $\{b_n\}$ are positive sequences. Sometimes, Always, Never:

   a) If $\lim\limits_{n\to\infty} \frac{a_n}{b_n} = 0$ and $\sum\limits_{k=1}^{\infty} a_k$ converges, then $\sum\limits_{k=1}^{\infty} b_k$ converges.

   b) If $\lim\limits_{n\to\infty} \frac{a_n}{b_n} = \infty$ and $\sum\limits_{k=1}^{\infty} b_n$ diverges, then $\sum\limits_{k=1}^{\infty} a_n$ diverges.

Sometimes. For $a_n = \frac{1}{n^3}$ and $b_n = \frac{1}{n^2}$ it is true. For $a_n = \frac{1}{n^2}$ and $b_n = \frac{1}{n}$ it is false.

Always.

B. If $\sum\limits_{n=1}^{\infty} a_n = s$ converges by the comparison Test using $t = \sum\limits_{n=1}^{\infty} b_n$, then
$$s - s_n \leq t - t_n.$$
This means that $t - t_n$ (which may be easier to calculate) is an upper estimate for $s - s_n$.

3) In question 1c), $\sum\limits_{n=1}^{\infty} \frac{1}{n+2^n}$ converges.

   a) Find s_4, the fourth partial sum.

$s_4 = \frac{1}{1+2} + \frac{1}{2+4} + \frac{1}{3+8} + \frac{1}{4+16}$

$= \frac{1}{3} + \frac{1}{6} + \frac{1}{11} + \frac{1}{20} = \frac{141}{220}$.

b) Estimate the difference between
    this series and its fourth partial
    sum.

From 1c), $b_n = \frac{1}{2^n}$ may be used to show

$\sum \frac{1}{n+2^n}$ converges. Let $s = \sum_{n=1}^{\infty} \frac{1}{n+2^n}$ and

$t = \sum_{n=1}^{\infty} \frac{1}{2^n}$. Then $s - s_4 \leq t - t_4$.

Since t is the result of a geometric series
we can calculate it:

$$t = \frac{\frac{1}{2}}{1-\frac{1}{2}} = 1$$

$$t_4 = \frac{1}{2} + \frac{1}{4} + \frac{1}{8} + \frac{1}{16} = \frac{31}{32}.$$

Thus $t - t_4 = 1 - \frac{31}{32} = \frac{1}{32}$ and

$s - s_4 \leq \frac{1}{32}$. Therefore we know $s_4 \left(\frac{141}{220} \right)$ is

within $\frac{1}{32}$ of the value of s.

Alternating Series

Concepts to Master

A. Alternating Series Tests
B. Estimating the sum of a convergent alternating series

Summary and Focus Questions

A. An <u>alternating series</u> has successive terms of opposite signs - that is, has either the form $\sum_{n=1}^{\infty} (-1)^n a_n$ or $\sum_{n=1}^{\infty} (-1)^{n+1} a_n$, where $a_n > 0$.

The <u>Alternating Series Test</u>:

> If $\{a_n\}$ is a decreasing sequence with $\lim_{n \to \infty} a_n = 0$, then $\sum_{n=1}^{\infty} (-1)^n a_n$ converges.

1) Is $\sum_{n=1}^{\infty} \frac{\sin n}{n}$ an alternating series?

No, although some terms are positive and others negative.

2) Determine whether each converge:

a) $\sum_{n=1}^{\infty} \frac{(-1)^n}{e^n}$.

This is an alternating series with $a_n = \frac{1}{e^n}$.
$\frac{1}{e^{n+1}} \leq \frac{1}{e^n}$, so the terms decrease.
Since $\lim_{n \to \infty} \frac{1}{e^n} = 0$, $\sum_{n=1}^{\infty} \frac{(-1)^n}{e^n}$ converges by the Alternating Series Test.

b) $\sum_{n=1}^{\infty} \frac{(-1)^n n}{n+1}$.

This is an alternating series but the other conditions for the test do not hold.
However, $\lim_{n \to \infty} \frac{(-1)^n n}{n+1} \neq 0$, so $\sum_{n=1}^{\infty} \frac{(-1)^n n}{n+1}$ diverges by the Divergence Test.

B. If $\sum_{n=1}^{\infty} (-1)^n a_n$ is an alternating series with $0 \le a_{n+1} \le a_n$ and $\lim_{n \to \infty} a_n = 0$ then

$$\left| s_n - s \right| \le a_{n+1}.$$

Since the difference between the limit of the series and the nth partial sum does not exceed a_{n+1}, s_n may be used to approximate s to within an accuracy of a_{n+1} by s_n.

3) The series $\sum_{n=1}^{\infty} \frac{(-1)^n}{\sqrt{n+1}}$ converges.

   Estimate the error between the sum of the series and its fifteenth partial sum.

Since the series alternates and $\frac{1}{\sqrt{n+1}}$ decreases to 0, the error $\left| s_{15} - s \right|$ does not exceed $a_{16} = \frac{1}{\sqrt{16+1}} = 0.2$.

4) For what value of n is s_n, the nth partial sum, within 0.001 of

   $$s = \sum_{n=1}^{\infty} \frac{(-1)^{n+1}}{2n+5}?$$

$\left| s_n - s \right| \le a_{n+1} = \frac{1}{(2n+1)+5} \le 0.001$

$2n + 6 \ge 1000$, $2n \ge 994$, $n \ge 497$.

Let $n = 497$. Then s_{497} is within 0.001 of s.

5) Approximate $\sum_{n=1}^{\infty} \frac{(-1)^{n+1}}{n^3}$ to within

   0.005.

$\left| s_n - s \right| \le a_{n+1} = \frac{1}{(n+1)^3} \le 0.005$

$(n+1)^3 \ge \frac{1}{0.005} = 200.$

For $n = 5$, $(n+1)^3 = 6^3 = 216 > 200.$

Therefore s_5 is within 0.005 of s.

$s_5 = \frac{1}{1^3} - \frac{1}{2^3} + \frac{1}{3^3} - \frac{1}{4^3} + \frac{1}{5^3}$

$= 1 - \frac{1}{8} + \frac{1}{27} - \frac{1}{64} + \frac{1}{125} \approx 0.9044.$

We do not know to what the series converges, but that value is within 0.005 of 0.9044.

Absolute Convergence and the Ratio and Root Tests

Concepts to Master

A. Absolute convergence; Conditional convergence
B. Ratio Test; Root Test

Summary and Focus Questions

A. For any series (with a_n not necessarily positive or alternating) the concept of convergence may be split into two subconcepts:

$$\sum_{k=1}^{\infty} a_k \text{ \underline{converges absolutely} means that } \sum_{k=1}^{\infty} |a_k| \text{ converges.}$$

$$\sum_{k=1}^{\infty} a_k \text{ \underline{converges conditionally} means that } \sum_{k=1}^{\infty} a_k \text{ converges and } \sum_{k=1}^{\infty} |a_k|$$
diverges.

Either one of absolute convergence or conditional convergence implies (ordinary) convergence. Conversely, convergence implies either absolute or conditional convergence. Thus *every* series must behave in exactly one of these three ways: diverge, converge absolutely, or converge conditionally.

1) Sometimes, Always, or Never:

   a) If $\sum_{n=1}^{\infty} |a_n|$ diverges, then $\sum_{n=1}^{\infty} a_n$ diverges.

   Sometimes. True for $a_n = n$ but false for $a_n = \frac{(-1)^n}{n}$.

b) If $\sum\limits_{n=1}^{\infty} a_n$ diverges, then

$\sum\limits_{n=1}^{\infty} |a_n|$ diverges.

Always.

c) If $a_n \geq 0$ for all n, then
$\sum\limits_{n=1}^{\infty} a_n$ is not conditionally
convergent.

Always.

2) Determine whether each converges
conditionally, converges absolutely, or
diverges.

a) $\sum\limits_{n=1}^{\infty} \frac{(-1)^n}{\sqrt{n}}$.

The series is alternating with $\frac{1}{\sqrt{n}}$
decreasing to 0 so it converges. It remains
to check for absolute convergence:
$\sum\limits_{n=1}^{\infty} \left| \frac{(-1)^n}{n} \right| = \sum\limits_{n=1}^{\infty} \frac{1}{\sqrt{n}}$ diverges $\left(p\text{-series with} \right.$
$\left. p = \frac{1}{2} \right)$. Thus $\sum\limits_{n=1}^{\infty} \frac{(-1)^n}{n}$ converges
conditionally.

b) $\sum\limits_{n=1}^{\infty} \frac{\sin n + \cos n}{n^3}$.

Check for absolute convergence first:
$\sum\limits_{n=1}^{\infty} \left| \frac{\sin n + \cos n}{n^3} \right| = \sum\limits_{n=1}^{\infty} \frac{|\sin n + \cos n|}{n^3}$.
Since $|\sin n + \cos n| \leq 2$ and $\sum\limits_{n=1}^{\infty} \frac{2}{n^3}$
converges (it is a p-series with $p = 3$),
$\sum\limits_{n=1}^{\infty} \frac{|\sin n + \cos n|}{n^3}$ converges by the Comparison
Test. Thus $\sum\limits_{n=1}^{\infty} \frac{\sin n + \cos n}{n^3}$ converges
absolutely.

3) Does $\sum\limits_{n=1}^{\infty} \frac{\sin e^n}{e^n}$ converge?

Yes. The series is not alternating but does contain both positive and negative terms. Check for absolute convergence:

$$\sum_{n=1}^{\infty} \left| \frac{\sin e^n}{e^n} \right| = \sum_{n=1}^{\infty} \frac{|\sin e^n|}{e^n}.$$

Since $\left| \sin e^n \right| \leq 1$, $\frac{|\sin e^n|}{e^n} \leq \frac{1}{e^n}$.

$\sum\limits_{n=1}^{\infty} \frac{1}{e^n}$ converges $\left(\text{geometric series with} \right.$

$r = \frac{1}{e} \Big)$ so by the Comparison Test

$\sum\limits_{n=1}^{\infty} \left| \frac{\sin e^n}{e^n} \right|$ converges. Therefore $\sum\limits_{n=1}^{\infty} \frac{\sin e^n}{e^n}$

converges absolutely and hence converges.

B. Let a_n be a sequence of nonzero terms. The following may be used to determine whether $\sum\limits_{n=1}^{\infty} a_n$ converges.

Ratio Test:

1. If $\lim\limits_{n\to\infty} \left| \frac{a_{n+1}}{a_n} \right| = L < 1$, then $\sum\limits_{n=1}^{\infty} a_n$ converges absolutely (and therefore converges).

2. If $\lim\limits_{n\to\infty} \left| \frac{a_{n+1}}{a_n} \right| = L > 1$ or $\lim\limits_{n\to\infty} \left| \frac{a_{n+1}}{a_n} \right| = \infty$, then $\sum\limits_{n=1}^{\infty} a_n$ diverges.

In case $\lim\limits_{n\to\infty} \left| \frac{a_{n+1}}{a_n} \right| = 1$, the Ratio Test fails - $\sum\limits_{n=1}^{\infty} a_n$ may converge or diverge.

Root Test:

1. If $\lim\limits_{n\to\infty} \sqrt[n]{|a_n|} = L < 1$, then $\sum\limits_{n=1}^{\infty} a_n$ converges absolutely (and therefore converges).

2. If $\lim\limits_{n\to\infty} \sqrt[n]{|a_n|} = L > 1$ or $\lim\limits_{n\to\infty} \sqrt[n]{|a_n|} = \infty$, then $\sum\limits_{n=1}^{\infty} a_n$ diverges.

If $\lim\limits_{n\to\infty} \sqrt[n]{|a_n|} = 1$, the Root Test fails - $\sum\limits_{n=1}^{\infty} a_n$ may converge or diverge.

The Ratio and Root Tests involve no other series or functions and so are relatively easy to use, but both fail for $L = 1$. The Ratio Test usually provides an answer when a_n contains exponentials and/or factorials or when a_n is defined by a recurrence relation. It will not work when a_n is a rational function of n. The Root Test works best when a_n contains an expression to the nth power.

4) Determine whether each converges by the Ratio Test:

a) $\displaystyle\sum_{n=1}^{\infty} \frac{n}{4^n}$.

$$\lim_{n\to\infty}\left|\frac{a_{n+1}}{a_n}\right| = \lim_{n\to\infty}\frac{\frac{n+1}{4^{n+1}}}{\frac{n}{4^n}} = \lim_{n\to\infty}\frac{n+1}{4n} = \frac{1}{4}.$$

Thus $\displaystyle\sum_{n=1}^{\infty}\frac{n}{4^n}$ converges absolutely and therefore converges.

b) $\displaystyle\sum_{n=1}^{\infty} \frac{(-4)^n}{n!}$.

$$\lim_{n\to\infty}\left|\frac{a_{n+1}}{a_n}\right| = \lim_{n\to\infty}\left|\frac{\frac{(-4)^{n+1}}{(n+1)!}}{\frac{(-4)^n}{n!}}\right| = \lim_{n\to\infty}\frac{4}{n+1} = 0.$$

Thus $\displaystyle\sum_{n=1}^{\infty}\frac{(-4)^n}{n!}$ converges absolutely and therefore converges.

c) $\displaystyle\sum_{n=1}^{\infty} \frac{(-1)^n}{\sqrt[3]{n}}$.

$$\lim_{n\to\infty}\left|\frac{a_{n+1}}{a_n}\right| = \lim_{n\to\infty}\left|\frac{\frac{(-1)^{n+1}}{\sqrt[3]{n+1}}}{\frac{(-1)^n}{\sqrt[3]{n}}}\right| = \lim_{n\to\infty}\sqrt[3]{\frac{n}{n+1}} = 1.$$

The Ratio Test <u>fails</u>. The Alternating Series Test can be used to show that this series converges.

d) $\displaystyle\sum_{n=1}^{\infty} a_n$, where $a_1 = 4$ and $a_{n+1} = \frac{3a_n}{2n+1}$.

$$\lim_{n\to\infty}\left|\frac{a_{n+1}}{a_n}\right| = \lim_{n\to\infty}\frac{\frac{3a_n}{2n+1}}{a_n} = \lim_{n\to\infty}\frac{3}{2n+1} = 0.$$

Thus $\displaystyle\sum_{n=1}^{\infty} a_n$ converges absolutely and therefore converges.

5) Determine whether each converges by the Root Test:

a) $\sum_{n=1}^{\infty} \frac{1}{(n+1)^n}$.

$\lim_{n\to\infty} \sqrt[n]{|a_n|} = \lim_{n\to\infty} \sqrt[n]{\frac{1}{(n+1)^n}} = \lim_{n\to\infty} \frac{1}{n+1} = 0.$

Thus $\sum_{n=1}^{\infty} \frac{1}{(n+1)^n}$ converges absolutely and therefore converges.

b) $\sum_{n=1}^{\infty} \frac{3^n}{n^3}$.

$\lim_{n\to\infty} \sqrt[n]{|a_n|} = \lim_{n\to\infty} \sqrt[n]{\frac{3^n}{n^3}} = \lim_{n\to\infty} \frac{3}{n^{3/n}} = \frac{3}{1} = 3.$

Therefore by the Root Test $\sum_{n=1}^{\infty} \frac{3^n}{n^3}$ diverges.

$\left(\lim_{n\to\infty} n^{3/n} = 1 \text{ using the methods of} \right.$
Section 3.9: Let $y = \lim_{n\to\infty} n^{3/n}$.
Then $\ln y = \lim_{n\to\infty} \ln n^{3/n} = \lim_{n\to\infty} \frac{3\ln n}{n}$
$\left(\text{form } \frac{\infty}{\infty}\right) = \lim_{n\to\infty} \frac{\frac{3}{n}}{1} = 0.$
Thus $y = e^0 = 1.\Big)$

c) $\sum_{n=1}^{\infty} \frac{(-1)^n}{n}$

$\lim_{n\to\infty} \sqrt[n]{\left|\frac{(-1)^n}{n}\right|} = \lim_{n\to\infty} \frac{1}{\sqrt[n]{n}} = 1.$

The Root Test fails, but we know this series converges because it is the alternating harmonic series.

Strategy for Testing Series

Concepts to Master

Applying all the previous series tests

Summary and Focus Questions

Different series lend themselves to different series tests. Follow the eight step strategy outlined in the text. Here is a brief summary of each test together with a representative example:

Test Name	Condition(s)	Conclusion(s)	Sample
Divergence	$\lim\limits_{n \to \infty} a_n \neq 0$	$\sum a_n$ diverges	$\sum\limits_{n=1}^{\infty} \frac{n}{n+1}$
Integral	Find $f(x)$, decreasing with $f(n) = a_n \geq 0$.	$\int_1^{\infty} f(x)dx$ converges if and only if $\sum a_n$ converges.	$\sum\limits_{n=1}^{\infty} \frac{\ln n}{n}$ $\int \frac{\ln x}{x}\, dx = \frac{(\ln x)^2}{2}$
p-series	$a_n = \frac{1}{n^p}$	$\sum \frac{1}{n^p}$ converges if and only if $p > 1$.	$\sum\limits_{n=1}^{\infty} \frac{1}{n^3}$
Geometric Series	$a_n = ar^{n-1}$	$\sum ar^{n-1}$ converges if and only if $\|r\| < 1$.	$\sum\limits_{n=1}^{\infty} \frac{2}{3^n}$
Alternating Series	$a_n = (-1)^n b_n$ $(b_n \geq 0)$	$\sum a_n$ converges if $\lim\limits_{n \to \infty} b_n = 0$ and b_n is decreasing.	$\sum\limits_{n=1}^{\infty} \frac{(-1)^n}{n}$

24

Comparison $0 \leq a_n \leq b_n$ If $\sum b_n$ converges $\sum\limits_{n=1}^{\infty} \frac{1}{n^3+1}$

then $\sum a_n$ converges.

If $\sum a_n$ diverges then   Choose $b_n = \frac{1}{n^3}$.

$\sum b_n$ diverges.

Limit $\lim\limits_{n\to\infty} \frac{a_n}{b_n} = c$ If $0 < c < \infty$, $\sum a_n$ $\sum\limits_{n=1}^{\infty} \frac{1}{2n^3+1}$

Comparison   converges if and only

if $\sum b_n$ does.   Choose $b_n = \frac{1}{n^3}$.

Ratio $\lim\limits_{n\to\infty} \left| \frac{a_{n+1}}{a_n} \right| = L$ If $L < 1$, $\sum a_n$ $\sum\limits_{n=1}^{\infty} \frac{n^2}{3^n}$

converges absolutely.

If $L > 1$, $\sum a_n$

diverges.

Root $\lim\limits_{n\to\infty} \sqrt[n]{|a_n|} = L$ If $L < 1$, $\sum a_n$ $\sum\limits_{n=1}^{\infty} \left(\frac{2n+1}{3n+4} \right)^n$

converges absolutely.

If $L > 1$, $\sum a_n$

diverges.

For each series, what is a good test to apply first for each?

1) $\sum\limits_{n=1}^{\infty} \frac{(n+1)^n}{4^n}$.

Root Test, because of the $(n+1)^n$ and 4^n exponentials.

2) $\sum\limits_{n=1}^{\infty} \frac{5}{n^2+6n+3}$.

Comparison Test, compare to $\sum \frac{1}{n^2}$.

3) $\sum\limits_{n=1}^{\infty} \frac{(-1)^n n}{n^6+1}$.

Alternating Series Test for convergence, then Comparison Test (compare to $\sum \frac{1}{n^5}$).

4) $\sum\limits_{n=1}^{\infty} \frac{2^n n^3}{n!}$.

Ratio Test, because of the $n!$ term.

Power Series

Concepts to Master

Power series in $(x - c)$; Interval of convergence; Radius of convergence

Summary and Focus Questions

A <u>power series in $(x - c)$</u> is an expression of the form $\sum\limits_{n=0}^{\infty} a_n(x - c)^n$, where a_n and c are constants. For any particular x, the value of the power series is an infinite series.

For some values of x it will converge while it may diverge for other values. The domain of a power series in $x - c$ is called its <u>interval of convergence</u> and contains those values of x for which the series converges. It consists of all real numbers from $c - R$ to $c + R$ for some R with $0 \leq R \leq \infty$. The number c is called the <u>center</u> and R is called the <u>radius</u> of convergence. When $R = 0$, the interval is the single point $\{c\}$. When $R = \infty$, the interval is $(-\infty, \infty)$.

The radius R is determined by applying the Ratio Test to $\sum\limits_{n=0}^{\infty} a_n(x - c)^n$.

When $0 < R < \infty$, the Ratio Test gives no information about the two endpoints $c - R$ and $c + R$. These endpoints may or may not be in the interval of convergence and must be tested separately using some other test. You should remember that the power series diverges for all x not in the interval of convergence.

1) True or False:
   A power series is an infinite series.

   False. A power series is an expression for a function, $f(x)$. For any x in the domain of f, $f(x)$ is a convergent series.

2) For what value of x does
    $\sum_{n=0}^{\infty} a_n(x-c)^n$ always converge?

At $x = c$, $\sum_{n=0}^{\infty} a_n(x-c)^n = a_0$. The power series always converges when x is the center.

3) Compute the value of:

   a) $\sum_{n=0}^{\infty} n^2(x-3)^n$ at $x = 4$.

When $x = 4$ the power series is
$\sum_{n=0}^{\infty} n^2 1^n = \sum_{n=0}^{\infty} n^2$. This infinite series diverges so there is no value for $x = 4$.

   b) $\sum_{n=0}^{\infty} 2^n(x-1)^n$ at $x = \frac{5}{6}$.

When $x = \frac{5}{6}$ the power series is
$\sum_{n=0}^{\infty} 2^n\left(-\frac{1}{6}\right)^n = \sum_{n=0}^{\infty}\left(-\frac{1}{3}\right)^n$
$= 1 - \frac{1}{3} + \frac{1}{9} - \frac{1}{27} + \ldots$ This is a geometric series with $a = 1$, $r = -\frac{1}{3}$. It converges to $\frac{1}{1-\left(-\frac{1}{3}\right)} = \frac{3}{4}$.

4) Find the center, radius, and interval of convergence for:

   a) $\sum_{n=1}^{\infty} (2n)!(x-1)^n$.

Use the Ratio Test:
$\lim_{n\to\infty}\left|\frac{(2(n+1))!(x-1)^{n+1}}{(2n)!(x-1)^n}\right|$
$= \lim_{n\to\infty}(2n+1)(2n+2)|x-1| = \infty$,
except when $x = 1$. Thus the center is 1, radius is 0, and interval of convergence is $\{1\}$.

   b) $\sum_{n=1}^{\infty} \frac{(x-4)^n}{2^n n}$.

Use the Ratio Test:
$\lim_{n\to\infty}\left|\frac{(x-4)^{n+1}}{2^{n+1}(n+1)} \cdot \frac{2^n n}{(x-4)^n}\right|$
$= \lim_{n\to\infty}\frac{n}{2(n+1)}|x-4| = \frac{1}{2}|x-4|$.
From $\frac{1}{2}|x-4| < 1$ we conclude $|x-4| < 2$. The center is 4 and radius is 2. When $|x-4| = 2$, $x = 2$ or $x = 6$. At $x = 2$, the power series becomes $\sum_{n=1}^{\infty} \frac{(-1)^n}{n}$ which converges by the Alternating Series Test. At $x = 6$ the power series is the divergent harmonic series $\sum_{n=1}^{\infty} \frac{1}{n}$. The interval of convergence is $[2, 6)$.

c) $\sum\limits_{n=1}^{\infty} \frac{x^n}{n!}$

Use the Ratio Test:

$$\lim_{n\to\infty} \left| \frac{\frac{x^{n+1}}{(n+1)!}}{\frac{x^n}{n!}} \right| = \lim_{n\to\infty} \frac{|x|}{n} = 0 \text{ for all values}$$

of x. Thus the interval of convergence is all real numbers, $(-\infty, \infty)$.

5) Suppose $\sum\limits_{n=0}^{\infty} a_n(x-5)^n$ converges for $x = 8$. For what other values must it converge?

The interval of convergence is $5 - R$ to $5 + R$.

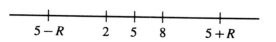

Because 8 is in this interval $8 \leq 5 + R$. Thus $3 \leq R$ which means the interval at least contains all numbers between $5 - 3$ and $5 + 3$. Thus the power series converges for at least all $x \in (2, 8]$.

Representation of Functions as Power Series

Concepts to Master

A. Representing a function by manipulating known series
B. Differentiation and integration of power series

Summary and Focus Questions

A. We will take a known function and write it as a power series. Start with a power series expression for a similar function and by substitution (such as x^2 for x) and multiplication by constants and powers of x turn the known power series into one for the given function.

Here is an example - find a power series for $\frac{2x}{1+x^2}$:

From $\frac{1}{1-x} = \sum\limits_{n=0}^{\infty} x^n$ substitute $-x^2$ for x to obtain

$\frac{1}{1+x^2} = \sum\limits_{n=0}^{\infty} (-x^2)^n = \sum\limits_{n=0}^{\infty} (-1)^n x^{2n}$ and then multiply by $2x$ to obtain

$\frac{2x}{1+x^2} = \sum\limits_{n=0}^{\infty} (-1)^n x^{2n}(2x) = \sum\limits_{n=0}^{\infty} (-1)^n 2x^{2n+1}$. The interval of convergence

for $\frac{1}{1-x}$ is $(-1, 1)$, so from $\left| x \right| < 1$, $\left| x^2 \right| < 1$ still yields $\left| x \right| < 1$.

1) Given $\frac{1}{1-x} = \sum\limits_{n=0}^{\infty} x^n$, $\left| x \right| < 1$, find a power series expression for:

a) $\frac{1}{2-x}$.

$\frac{1}{2-x} = \frac{1}{2\left(1-\frac{x}{2}\right)} = \frac{1}{2} \cdot \frac{1}{1-\frac{x}{2}}$. Thus

$\frac{1}{2-x} = \frac{1}{2}\sum\limits_{n=0}^{\infty} \left(\frac{x}{2}\right)^n = \sum\limits_{n=0}^{\infty} \frac{1}{2} \cdot \frac{x^n}{2^n}$

$= \sum\limits_{n=0}^{\infty} \frac{x^n}{2^{n+1}}$.

From $\left| \frac{x}{2} \right| < 1$, $\left| x \right| < 2$, so $(-2, 2)$ is the interval of convergence.

b) $\frac{x}{x^3-1}$.

$\frac{x}{x^3-1} = -\frac{x}{1-x^3}$.

First substitute x^3 for x:

$$\frac{1}{1-x^3} = \sum_{n=0}^{\infty}(x^3)^n = \sum_{n=0}^{\infty}x^{3n}.$$

Now multiply by $-x$:

$$\frac{-x}{1-x^3} = -x\sum_{n=0}^{\infty}x^{3n} = -\sum_{n=0}^{\infty}x^{3n+1}.$$

Originally $|x| < 1$, hence $|x^3| < 1$. Thus the interval of convergence is still $(-1, 1)$.

B. If $f(x) = \sum_{n=0}^{\infty}a_n(x-c)^n$ has interval of convergence with radius R then:

1. f is continuous on $(c - R, c + R)$.

2. For all $x \in (c - R, c + R)$, f may be differentiated term by term:
$$f'(x) = \sum_{n=1}^{\infty}na_n(x-c)^{n-1}.$$

3. For all $x \in (c - R, c + R)$, f may be integrated term by term:
$$\int f(x)dx = C + \sum_{n=0}^{\infty}\frac{a_n(x-c)^{n+1}}{n+1}.$$

Both $f'(x)$ and $\int f(x)dx$ have radius of convergence R but the endpoints $c - R$ and $c + R$ must still be checked individually.

Term by term integration and differentiation may be used to find more power series for known functions.

For example, to find a power series for $\ln(1 + x^2)$ recall from part A that the series for $\frac{1}{1-x}$ is used to find that

$$\frac{2x}{1+x^2} = \sum_{n=0}^{\infty}(-1)^2 2x^{2n+1}.$$

Thus $\int \frac{2x}{1+x^2}\, dx = \int\sum_{n=0}^{\infty}(-1)^n 2x^{2n+1}\, dx = \sum_{n=0}^{\infty}(-1)^n 2\int x^{2n+1}\, dx$

or $\ln(1 + x^2) = \sum_{n=0}^{\infty}(-1)^n 2\frac{x^{2n+2}}{2n+2} = \sum_{n=0}^{\infty}(-1)^n \frac{x^{2n+2}}{2n+1}$.

We still have $|x| < 1$ but need to check $x = 1$ and $x = -1$. At both $x = 1$ and $x = -1$, $x^{2n+2} = 1$. $\sum\limits_{n=0}^{\infty} \frac{(-1)^n}{2n+1}$ converges so the interval of convergence is $[-1,\ 1]$.

2) Find $f'(x)$ for
$$f(x) = \sum_{n=0}^{\infty} 2^n (x-1)^n.$$

$$f'(x) = \sum_{n=1}^{\infty} 2^n n (x-1)^{n-1}.$$

3) Find $\int f(x)dx$ where
$$f(x) = \sum_{n=0}^{\infty} \frac{(x-3)^n}{n!}.$$

$$\int f(x)dx = C + \sum_{n=0}^{\infty} \frac{1}{n!} \frac{(x-3)^{n+1}}{n+1}$$
$$= C + \sum_{n=0}^{\infty} \frac{(x-3)^{n+1}}{(n+1)!}.$$

4) Is $f(x) = \sum\limits_{n=0}^{\infty} \frac{x^n}{4^n}$ continuous at $x = 2$?

Yes. 2 is in $[-4,\ 4]$, the interval of convergence for $f(x)$.

5) Given $\frac{1}{1-x} = \sum\limits_{n=0}^{\infty} x^n$, $|x| < 1$, find a power series for $\frac{1}{(1+x)^2}$.

Substitute $-x$ for x:
$$\frac{1}{1+x} = \sum_{n=0}^{\infty} (-x)^n = \sum_{n=0}^{\infty} (-1)^n x^n.$$
Differentiate term by term:
$$\frac{-1}{(1+x)^2} = \sum_{n=0}^{\infty} (-1)^n n x^{n-1}.$$
Thus $\frac{1}{(1+x)^2} = (-1)\sum\limits_{n=0}^{\infty} (-1)^n n x^{n-1}$
$$= \sum_{n=0}^{\infty} (-1)^{n+1} n x^{n-1}.$$
$|-x| < 1$ is the same as $|x| < 1$, the radius of convergence is $R = 1$. At $x = 1$ we have $\sum\limits_{n=0}^{\infty} (-1)^{n+1} n$ and at $x = -1$ we have
$$\sum_{n=0}^{\infty} (-1)^{n+1} n (-1)^{n-1} = \sum_{n=0}^{\infty} n.$$ Both series diverge so the interval of convergence is $(-1,\ 1)$.

6) Find $\int \frac{1}{1+x^5} \, dx$ as a power series.

From $\frac{1}{1-x} = \sum_{n=0}^{\infty} x^n$,

$\frac{1}{1+x} = \sum_{n=0}^{\infty} (-x)^n = \sum_{n=0}^{\infty} (-1)^n x^n$.

Thus $\frac{1}{1+x^5} = \sum_{n=0}^{\infty} (-1)^n x^{5n}$.

$\int \frac{1}{1+x^5} \, dx = \sum_{n=0}^{\infty} \int (-1)^n x^{5n} \, dx$

$= \sum_{n=0}^{\infty} \frac{(-1)^n}{5n+1} x^{5n+1} + C$.

Taylor and Maclaurin Series

Concepts to Master

A. Taylor series; Maclaurin series; Analytic functions
B. Taylor polynomials of degree n
C. Taylor's Formula with Remainder (Lagrange's form)
D. Multiplication and division of power series

Summary and Focus Questions

A. If a function $y = f(x)$ can be expressed as a power series it will be in this form, called the <u>Taylor series of f at c</u>:

$$f(x) = \sum_{n=0}^{\infty} \frac{f^{(n)}(C)}{n!}(x - c)^n$$

$$= f(c) + f'(c)(x - c) + \frac{f''(c)}{2!}(x - c)^2 + \ldots + \frac{f^{(n)}(c)}{n!}(x - c)^n + \ldots$$

In the special case of $c = 0$, this is called the <u>Maclaurin series of f</u>.

Not all functions can be represented by their power series. If f can be represented as a power series about c we say f is <u>analytic at c</u>. In such a case f is equal to its Taylor series at c.

Here are some basic Maclaurin series:

$$e^x = \sum_{n=0}^{\infty} \frac{x^n}{n!} \text{ for all } x. \qquad \ln(1 + x) = \sum_{n=0}^{\infty} \frac{(-1)^{n+1}}{n} x^n \text{ for } x \in (-1,\, 1].$$

$$\sin x = \sum_{n=0}^{\infty} \frac{(-1)^n}{(2n+1)!} x^{2n+1} \text{ for all } x. \qquad \cos x = \sum_{n=0}^{\infty} \frac{(-1)^n}{(2n)!} x^{2n} \text{ for all } x.$$

$$\frac{1}{1-x} = \sum_{n=0}^{\infty} x^n \text{ for } |x| < 1. \qquad \tan^{-1}x = \sum_{n=0}^{\infty} (-1)^n \frac{x^{2n+1}}{2n+1} \text{ for } |x| \le 1.$$

33

To find a Taylor series for a given $y = f(x)$, you must find a formula for $f^{(n)}(c)$ usually in terms of n. Computing the first few derivatives $f'(c)$, $f''(c)$, $f'''(c)$, $f^{(4)}(c), \ldots$ often helps.

For some functions f it is easier to find the Taylor series for $f'(x)$ or $\int f(x)\,dx$ then integrate or differentiate term by term to obtain the series for f. In other cases, substitutions in the basic Maclaurin series can be used to find the Taylor series for f.

1) Find the Taylor series for $\frac{1}{x}$ at $c = 1$.

   a) directly from the definition.

We must find the general form of $f^{(n)}(1)$:

$f(x) = x^{-1}$	$f(1) = 1$
$f'(x) = -x^{-2}$	$f'(1) = -1$
$f''(x) = 2x^{-3}$	$f''(1) = 2$
$f'''(x) = -6x^{-4}$	$f'''(1) = -6$
$f^{(4)}(x) = 24x^{-5}$	$f^{(4)}(1) = 24$

In general, $f^{(n)}(1) = (-1)^n n!$
Thus the Taylor series is
$$\sum_{n=0}^{\infty} \frac{(-1)^n n!}{n!}(x-1)^n = \sum_{n=0}^{\infty} (-1)^n (x-1)^n$$
$$= \sum_{n=0}^{\infty} (1-x)^n.$$

   b) using substitution in a geometric series $\left(\text{Hint: } \frac{1}{x} = \frac{1}{1-(1-x)}\right)$.

In the form $\frac{1}{1-(1-x)}$, this is the value of a geometric series with $a = 1$, $r = 1 - x$.
Thus $\frac{1}{x} = \sum_{n=1}^{\infty} 1(1-x)^{n-1} = \sum_{n=0}^{\infty} (1-x)^n$.

   c) What is the interval of convergence for your answer to part a)?

$$\lim_{n\to\infty} \left| \frac{(1-x)^{n+1}}{(1-x)^n} \right| = \lim_{n\to\infty} |1-x| = |1-x|.$$
$|1-x| < 1$ is equivalent to $0 < x < 2$.
At $x = 0$ and $x = 2$ the terms of $\sum_{n=0}^{\infty} (1-x)^n$ do not approach zero so the series diverges. The interval of convergence is $(0, 2)$.

2) Find directly the Maclaurin series for
$f(x) = e^{4x}$.

$$f(x) = e^{4x} \qquad f(0) = 1$$
$$f'(x) = 4e^{4x} \qquad f'(0) = 4$$
$$f''(x) = 16e^{4x} \qquad f''(0) = 16$$
In general $f^{(n)}(0) = 4^n$.

Thus $e^{4x} = \sum_{n=0}^{\infty} \frac{4^n}{n!} x^n$.

3) True, False:
If f is analytic at c, then
$$f(x) = \sum_{n=0}^{\infty} \frac{f^{(n)}(c)}{n!} (x - c)^n.$$

True.

4) Obtain the Maclaurin series for $\frac{x}{1+x^2}$
from the series for $\frac{1}{1-x} = \sum_{n=0}^{\infty} x^n$.

$\frac{1}{1-x} = 1 + x + x^2 + \ldots + x^n + \ldots$
Substitute $-x^2$ for x:
$\frac{1}{1+x^2} = 1 - x^2 + x^4 + \ldots$
$\qquad + (-1)^n x^{2n} + \ldots$
Now multiply by x:
$\frac{x}{1+x^2} = x - x^3 + x^5 + \ldots$
$\qquad + (-1)^n x^{2n+1} + \ldots$
$\qquad = \sum_{n=0}^{\infty} (-1)^n x^{2n+1}$

5) Using $e^x = \sum_{n=0}^{\infty} \frac{x^n}{n!}$ find the Maclaurin
series for $\sinh x$.
(Remember, $\sinh x = \frac{e^x - e^{-x}}{2}$.)

$e^x = 1 + x + \frac{x^2}{2!} + \ldots + \frac{x^n}{n!} + \ldots$
Thus $e^{-x} = 1 - x + \frac{x^2}{2!} + \ldots$
$\qquad + \frac{(-1)^n x^n}{n!} + \ldots$
and subtracting term by term (the even terms cancel)
$e^x - e^{-x} = 2x + \frac{2x^3}{3!} + \ldots + \frac{2x^{2n+1}}{(2n+1)!} + \ldots$
Thus $\sinh x = x + \frac{x^3}{3!} + \ldots$
$\qquad + \frac{x^{2n+1}}{(2n+1)!} + \ldots = \sum_{n=0}^{\infty} \frac{x^{2n+1}}{(2n+1)!}$.

6) Find the infinite series expression for
$\int_0^{1/2} \frac{1}{1-x^3} \, dx$.

$\frac{1}{1-x} = \sum_{n=0}^{\infty} x^n$. Thus $\frac{1}{1-x^3} = \sum_{n=0}^{\infty} x^{3n}$.

$\int \frac{1}{1-x^3} \, dx = \int \sum_{n=0}^{\infty} x^{3n} \, dx = \sum_{n=0}^{\infty} \frac{x^{3n+1}}{3n+1} F(x)$.
$\int_0^{1/2} \frac{1}{1-x^3} \, dx = F\left(\frac{1}{2}\right) - F(0)$

$\qquad = \sum_{n=0}^{\infty} \frac{\left(\frac{1}{2}\right)^{3n+1}}{3n+1} - 0 = \sum_{n=0}^{\infty} \frac{1}{(6n+2)8^n}$.

B. If $f^{(n)}(c)$ exists, the <u>Taylor polynomial of degree n for f about c</u> is

$$T_n(x) = f(c) + \frac{f'(c)}{1!}(x-c) + \frac{f''(c)}{2!}(x-c)^2 + \ldots \frac{f^{(n)}(c)}{n!}(x-c)^n.$$

$T_n(x)$ is simply the nth partial sum of the Taylor series for $f(x)$.

$T_1(x) = f(c) + f'(c)(x-c)$ is the familiar equation for the tangent line to $y = f(x)$ at $x = c$. $T_2(x)$ the "tangent parabola."

Since the first n derivatives of T_n at c are equal to the corresponding first n derivatives of f at c, $T_n(x)$ is an approximation for $f(x)$ if x is near c. Recall that we used $T_1(x)$ and $T_2(x)$ for approximations in Chapter 2.

7) Let $f(x) = \sqrt{x}$.

    a) Construct the Taylor polynomial of degree 3 for $f(x)$ about $x = 1$.

$$f(x) = x^{1/2} \qquad f(1) = 1$$
$$f'(x) = \tfrac{1}{2}x^{-1/2} \qquad f'(1) = \tfrac{1}{2}$$
$$f''(x) = -\tfrac{1}{4}x^{-3/2} \qquad f''(1) = -\tfrac{1}{4}$$
$$f^{(3)}(x) = \tfrac{3}{8}x^{-5/2} \qquad f^{(3)}(1) = \tfrac{3}{8}$$
$$T_3(x)$$
$$= 1 + \tfrac{1}{2}(x-1) - \tfrac{1/4}{2!}(x-1)^2 + \tfrac{3/8}{3!}(x-1)^3$$
$$= 1 + \tfrac{1}{2}(x-1) - \tfrac{1}{8}(x-1)^2 + \tfrac{1}{16}(x-1)^3$$

    b) Approximate $\sqrt{\tfrac{3}{2}}$ using the Taylor polynomial of degree 3 from part a).

$$T_3\left(\tfrac{3}{2}\right) = 1 + \tfrac{1}{2}\left(\tfrac{1}{2}\right) - \tfrac{1}{8}\left(\tfrac{1}{2}\right)^2 + \tfrac{1}{16}\left(\tfrac{1}{2}\right)^3$$
$$= \tfrac{79}{64} \approx 1.2344.$$
$$\left(\text{To 4 decimals } \sqrt{\tfrac{3}{2}} = 1.2247.\right)$$

8) a) Find the Taylor polynomial of degree 6 for $f(x) = \cos x$ about $x = 0$.

$$f(0) = \cos 0 = 1$$
$$f'(x) = -\sin x \qquad f'(0) = 0$$
$$f''(x) = -\cos x \qquad f''(0) = -1$$
$$f^{(3)}(x) = \sin x \qquad f^{(3)}(0) = 0$$
$$f^{(4)}(x) = \cos x \qquad f^{(4)}(0) = 1$$
$$f^{(5)}(0) = 0 \text{ and } f^{(6)}(0) = -1$$
Thus $T_6(x) = 1 + \tfrac{-1}{2!}x^2 + \tfrac{1}{4!}x^4 - \tfrac{1}{6!}x^6$
$$T_6(x) = 1 - \tfrac{x^2}{2} + \tfrac{x^4}{24} - \tfrac{x^6}{720}.$$
Note, of course, that this is the first few terms of the Maclaurin series for cosine.

b) Use your answer to part a) to
approximate cos 1.

$$\cos 1 \approx T_6(1) = 1 - \frac{(1)^2}{2} + \frac{(1)^4}{24} - \frac{(1)^6}{720}$$
$$= 1 - 0.5 + 0.041667 - 0.001389$$
$$= 0.540278$$

(Note: cos 1 = 0.540302 to 6 decimal places.)

C. Taylor's Formula with Remainder (Lagrange's form):

Let f be a function whose $n + 1$ derivatives exist and are continuous in an interval I containing c. Then for $x \in I$,
$$f(x) = T_n(x) + R_n(x),$$
$$R_n(x) = \frac{f^{(n+1)}(z)}{(n+1)!}(x - c)^{n+1} \text{ for some number } z \text{ between } c \text{ and } x.$$

$R_n(x)$ is expressed in <u>Lagrange's form</u> for the remainder and is the <u>error</u> in approximating $f(x)$ with $T_n(x)$.

If f has derivatives of all orders and $\lim\limits_{n \to \infty} R_n(x) = 0$, then $f(x)$ is equal to its Taylor series on its interval of convergence. Thus the approximation $T_n(x)$ often can be made accurate to within any given number ε by selecting n such that $|R_n(x)| < \varepsilon$. This usually requires finding upper bound estimates for $|f^{(n+1)}(z)|$.

9) True or False:
   In general, the larger the number n, the closer the Taylor polynomial approximation is to the actual functional value.

   True.

10) a) Write Taylor's Formula with Remainder with $n = 3$ for $f(x) = x^{5/2}$ about $c = 4$.

   $f(x) = x^{5/2}$ $f(4) = 32$

   $f'(x) = \frac{5}{2}x^{3/2}$ $f'(4) = 20$

   $f''(x) = \frac{15}{4}x^{1/2}$ $f''(4) = \frac{15}{2}$

   $f'''(x) = \frac{15}{8}x^{-1/2}$ $f'''(4) = \frac{15}{16}$

$$T_3(x) = 32 + 20(x - 4) + \frac{15/2}{2!}(x - 4)^2$$
$$+ \frac{15/16}{3!}(x - 4)^3$$
$$= 32 + 20(x - 4) + \frac{15}{4}(x - 4)^2$$
$$+ \frac{15}{32}(x - 4)^3.$$

Since $f^{(4)}(x) = -\frac{15}{16}x^{-3/2}$,

$$R_3(x) = \frac{-(15/16)z^{-3/2}}{4!}(x - 4)^4. \text{ Therefore}$$

$$R_3(x) = \frac{-5}{128z^{3/2}}(x - 4)^4, \text{ for } z \text{ between } c$$
and x. Taylor's Formula is
$f(x) = T_3(x) + R_3(x)$, where $T_3(x)$ and
$R_3(x)$ are given above.

b) Estimate the error between
$f(5) = 5^{5/2}$ and $T_3(5)$.

An estimate of the error is

$$|R_3(5)| = \frac{5}{128z^{3/2}}(5 - 4)^4 = \frac{5}{128z^{3/2}}.$$

Since $4 < z < 5$, $8 = 4^{3/2} < z^{3/2}$.
Thus $|R_3(5)| < \frac{5}{128(8)} = \frac{5}{1024} \approx 0.00488$.

c) Compute $T_3(5)$.

$$T_3(5) = 32 + 20(1)^1 + \frac{15}{4}(1)^2 + \frac{5}{32}(1)^3$$
$$= \frac{1789}{32} \approx 55.9063.$$

d) To four decimal places, what is
the actual error?

To four decimals $f(5) = 5^{5/2} = 55.9017$.
Thus the error is
$55.9063 - 55.9017 = 0.0046$, which is
within 0.00488.

11) Find $\int_0^1 xe^{-x} \, dx$ using series to within
0.01.

$$\left(\text{We could use integration by parts:} \right.$$
$$\left. \int_0^1 xe^{-x} \, dx = (-xe^{-x} - e^{-x})\Big|_0^1 = 1 - \frac{2}{e}. \right)$$

$$e^x = \sum_{n=0}^{\infty} \frac{x^n}{n!}, \quad e^{-x} = \sum_{n=0}^{\infty} \frac{(-1)^n x^n}{n!},$$

$$\text{so } xe^{-x} = \sum_{n=0}^{\infty} \frac{(-1)^n x^{n+1}}{n!}. \text{ Thus}$$

$$\int_0^1 xe^{-x} \, dx = \int_0^1 \sum_{n=0}^{\infty} \frac{(-1)^n x^{n+1}}{n!} \, dx$$

$$= \sum_{n=0}^{\infty} \frac{(-1)^n x^{n+2}}{n!(n+2)} \Big|_0^1 = \sum_{n=0}^{\infty} \frac{(-1)^n}{n!(n+2)} - 0$$

$$= \sum_{n=0}^{\infty} \frac{(-1)^n}{n!(n+2)}$$

$$= \frac{1}{0!2} - \frac{1}{1!3} + \frac{1}{2!4} - \frac{1}{3!5} + \frac{1}{4!6} - \frac{1}{5!}$$

$$= \frac{1}{2} - \frac{1}{3} + \frac{1}{8} - \frac{1}{30} + \frac{1}{144}.$$

This is an alternating series and the fifth term $\frac{1}{144}$ is less than 0.01. Thus we may use $\int_0^1 xe^{-x}\,dx \approx \frac{1}{2} - \frac{1}{3} + \frac{1}{8} - \frac{1}{30} \approx 0.2583$. This is within 0.01 of the actual value $1 - \frac{2}{e} \approx 0.2642$. We note in passing that the series $\sum_{n=0}^{\infty} \frac{(-1)^n}{n!(n+2)} = 1 - \frac{2}{e}$.

D. Convergent series may be multiplied and divided like polynomials. Often finding the first few terms of the result is sufficient.

12) Find the first three terms of the Maclaurin series for $e^x \sin x$.

$e^x = 1 + x + \frac{x^2}{2} + \frac{x^3}{6} + \ldots$

$\sin x = x - \frac{x^3}{3!} + \frac{x^5}{5!} - \ldots$

Thus

$e^x \sin x = x\left(1 + x + \frac{x^2}{2} + \frac{x^3}{6} + \ldots\right)$
$\qquad - \frac{x^3}{6}\left(1 + x + \frac{x^2}{2} + \frac{x^3}{6} + \ldots\right)$
$\qquad + \frac{x^5}{120}\left(1 + x + \frac{x^2}{2} + \frac{x^3}{6} + \ldots\right)$
$\qquad + \ldots$

$= \left(x + x^2 + \frac{x^3}{2} + \frac{x^4}{6} + \ldots\right)$
$\qquad - \left(\frac{x^3}{6} + \frac{x^4}{6} + \frac{x^5}{12} + \frac{x^6}{36} + \ldots\right)$
$\qquad + \left(\frac{x^5}{120} + \frac{x^6}{120} + \frac{x^7}{240} + \ldots\right)$

$= x + x^2 + \frac{x^3}{3} + \ldots$

The Binomial Series

Concepts to Master

Binomial coefficients; Binomial series for $(1 + x)^k$

Summary and Focus Questions

For any real number k and $|x| < 1$, the <u>binomial series for $(1 + x)^k$</u> is the Maclaurin series for $(1 + x)^k$:

$$(1 + x)^k = \sum_{n=0}^{\infty} \binom{k}{n} x^n = 1 + kx + \frac{k(k-1)}{2} x^2 + \ldots + \binom{k}{n} x^n + \ldots$$

Here $\binom{k}{n}$ is the binomial coefficient, which is defined as $\binom{k}{n} = \frac{k(k-1)\ldots(k-n+1)}{n!}$, $n \geq 1$. Also $\binom{k}{0} = 1$ for all k.

The binomial series converges to $(1 + x)^k$ for these cases:

Condition on k	Interval of Convergence
$k \leq -1$	$(-1, 1)$
$-1 < k < 0$	$(-1, 1]$
k nonnegative integer	$(-\infty, \infty)$
$k > 0$, but not an integer	$[-1, 1]$

1) Find $\binom{4/3}{5}$.

$$\binom{4/3}{5} = \frac{\left(\frac{4}{3}\right)\left(\frac{1}{3}\right)\left(-\frac{2}{3}\right)\left(-\frac{5}{3}\right)\left(-\frac{8}{3}\right)}{5 \cdot 4 \cdot 3 \cdot 2 \cdot 1}$$
$$= \frac{-8}{3^6} = \frac{-8}{729}.$$

2) True or False:
   If k is a nonnegative integer, then the binomial series for $(1 + x)^k$ is a finite sum.

True.

3) Find the binomial series for $\sqrt[3]{1 + x}$.

$$\sum_{n=0}^{\infty} \binom{1/3}{n} x^n.$$

40

4) Find a power series for
$f(x) = \sqrt{4 + x}$.

$$\sqrt{4 + x} = 2\left(1 + \tfrac{x}{2}\right)^{1/2} = 2\sum_{n=0}^{\infty} \binom{\frac{1}{2}}{n}\left(\tfrac{x}{2}\right)^n$$
$$= \sum_{n=0}^{\infty} \binom{\frac{1}{2}}{n}\frac{x^n}{2^{n-1}}.$$

5) Find an infinite series for $\sin^{-1} x$.
$\left(\text{Hint: Start with } (1 + x)^{-1/2}.\right)$

$$\frac{1}{\sqrt{1+x}} = (1 + x)^{-1/2} = \sum_{n=0}^{\infty} \binom{-\frac{1}{2}}{n} x^n.$$
$$\frac{1}{\sqrt{1-x}} = \sum_{n=0}^{\infty} \binom{-\frac{1}{2}}{n} (-x)^n$$
$$= \sum_{n=0}^{\infty} (-1)^n \binom{-\frac{1}{2}}{n} x^n.$$
$$\frac{1}{\sqrt{1-x^2}} = \sum_{n=0}^{\infty} (-1)^n \binom{-\frac{1}{2}}{n} x^{2n}.$$
Now integrate term by term:
$$\sin^{-1} x = \sum_{n=0}^{\infty} \frac{(-1)^n \binom{-\frac{1}{2}}{n} x^{2n+1}}{2n+1}.$$

Section 10.12

Applications of Taylor Polynomials

Concepts to Master

Approximating function values with Taylor polynomials; Estimating error in approximations

Summary and Focus Questions

The nth degree Taylor polynomial for a function $f(x)$ at c,
$$T_n(x) = f(c) + f'(c)(x - c) + \frac{f''(c)}{2!}(x - c)^2 + \ldots + \frac{f^{(n)}(c)}{n!}(x - c)^n,$$
may be used to approximate $f(x)$ for any given x in the radius of convergence of the Taylor series for f. How good the approximation $f(x) \approx T_n(x)$ will be dependent on:

1. n - the larger n is, the better the estimate, and
2. x and c - the closer x is to c, the better the estimate.

This is shown in the remainder $R_n = f(x) - T_n(x)$ by Taylor's formula:
$$R_n(x) = \frac{f^{(n+1)}(z)}{(n+1)!}(x - c)^{n+1} \text{ where } z \text{ is between } x \text{ and } c.$$

$|R_n(x)|$ may be calculated or a graphing calculator can be used to estimate it. Or, in the special case where the Taylor series is an Alternating Series, $|R_n|$ may be estimated with $\left| \frac{f^{n+1}(x)}{(n+1)!}(x - c)^{n+1} \right|$, the $(n + 1)$st term of the Taylor series.

1) a) For $f(x) = \frac{1}{3+x}$, find $T_3(x)$ where $c = 2$.

$f(x) = (3 + x)^{-1}$ $f(2) = \frac{1}{5}$

$f'(x) = -(3 + x)^{-2}$ $f'(2) = \frac{-1}{25}$

$f''(x) = 2(3 + x)^{-3}$ $f''(2) = \frac{2}{125}$

$f'''(x) = -6(3 + x)^{-4}$ $f'''(2) = \frac{-6}{625}$

$T_3(x) = \frac{1}{5} - \frac{1}{25}(x - 2) + \frac{2/125}{2!}(x - 2)^2$

$\qquad - \frac{6/625}{3!}(x - 2)^3$

$\qquad = \frac{1}{5} - \frac{1}{25}(x - 2) + \frac{1}{125}(x - 2)^3$

$\qquad - \frac{1}{625}(x - 2)^3$

b) Use Taylor's formula to estimate the accuracy of $f(x) \approx T_3(x)$ when $1 \leq x \leq 4$.

From part a), $f^{(4)}(x) = 24(3+x)^{-5}$.
$|f(x) - T_3(x)| = |R_3(x)|$
$= \left| \frac{24(3+z)^{-5}}{4!}(x-2)^4 \right| = \frac{(x-2)^4}{(3+z)^5}$
where z is between x and 2.
From $1 \leq x \leq 4$, $-1 \leq x - 2 \leq 2$.
Thus $(x-2)^4 \leq 2^4 = 16$.
z is between 2 and x, so $1 \leq z \leq 4$,
$4 \leq 3 + z \leq 7$. Thus $4^5 \leq (3+z)^5 \leq 7^5$,
or $\frac{1}{4^5} \geq \frac{1}{(3+z)^5} \geq \frac{1}{7^5}$. Therefore
$|R_3(x)| = \frac{(x-2)^4}{(3+z)^5} \leq \frac{16}{4^5} = \frac{1}{64} \approx 0.0156$.

c) Check the accuracy of the estimate when $x = 3$.

$f(3) = \frac{1}{3+3} = \frac{1}{6} \approx 0.1667$
$T_3(3) = \frac{1}{5} - \frac{1}{25} + \frac{1}{125} - \frac{1}{625} = 0.1664$
Indeed $|f(3) - T_3(3)| < 0.0156$.

2) a) What degree Maclaurin polynomial is needed to approximate $\cos 0.5$ accurate to within 0.00001?

Rephrased, the question asks: For what n is $|R_n(0.5)| < 0.00001$ when $f(x) = \cos x$ and $c = 0$?
$|R_n(0.5)| = \left| \frac{f^{(n+1)}(z)}{(n+1)!}(0.5 - 0)^{n+1} \right|$
$= \frac{\left| f^{(n+1)}(z) \right| (0.5)^{n+1}}{(n+1)!} \leq \frac{(0.5)^{n+1}}{(n+1)!}$
since $\left| f^{(n+1)}(z) \right| \leq 1$.
Now $\frac{(0.5)^{n+1}}{(n+1)!} \geq 0.00001$ for $n = 1, 2, 3, 4$.
However, for $n = 5$,
$\frac{(0.5)^{n+1}}{(n+1)!} = 0.000022 < 0.00001$.
Thus the degree should be at least 5.

b) Use your answer to part a) to estimate $\cos 0.5$ to within 0.00001.

For $f(x) = \cos x$, $f(0) = 1$,
$f'(0) = 0$, $f''(0) = -1$,
$f^{(3)}(0) = 0$, $f^{(4)}(0) = 1$, and
$f^{(5)}(0) = 0$.
Thus $T_5(x) = 1 - \frac{x}{2!} + \frac{x}{4!}$ and
$T_5(0.5) \approx 0.87760$.

11

Three-Dimensional Analytic Geometry and Vectors

Cartoons courtesy of Sidney Harris. Used by permission.

Three-Dimensional Coordinate Systems

Concepts to Master

A. Rectangular three-dimensional coordinates; Distance between two points
B. Planes; Spheres

Summary and Focus Questions

A.

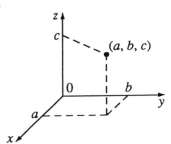

With the x, y, and z axes perpendicular to each other in three dimensional space, each triple (a, b, c) of reals corresponds to a unique point in space.

The distance between points P_1: (x_1, y_1, z_1) and P_2: (x_2, y_2, z_2) is

$$|P_1 P_2| = \sqrt{(x_2 - x_1)^2 + (y_2 - y_1)^2 + (z_2 - z_1)^2}.$$

1) Plot the points A: $(0, 5, 0)$
   B: $(5, 4, 6)$ C: $(1, -1, 3)$.

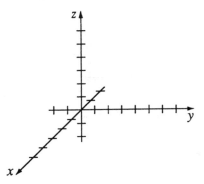

2) The coordinates of the points pictured
    are _____.

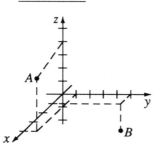

3) The distance between the points
    $(7, 5, -2)$ and $(4, 6, 1)$ is _____.

A: $(4, 1, 4)$ B: $(1, 5, -2)$

$$\sqrt{(7-4)^2 + (5-6)^2 + (-2-1)^2} = \sqrt{19}$$

B. Some simple linear equations in x, y, z with one or more of the variables
missing represent planes parallel to the axes of the missing variable. Only a
portion of each plane is drawn here:

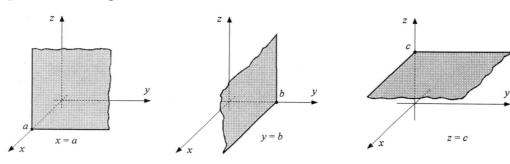

The equation of the sphere with
center (a, b, c) and radius R is
$(x - a)^2 + (y - b)^2 + (z - c)^2$
$\quad = R^2.$

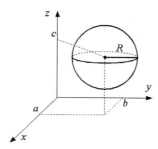

4) Describe the coordinates of each of points:

a)

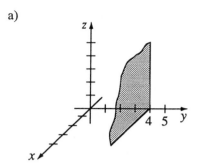

A point in the plane in the figure may have any values for x and z coordinates but must have a y coordinate of 4. The plane is all the points (x, y, z) such that $y = 4$.

b)

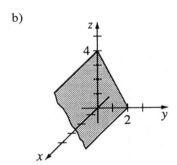

The plane is parallel to the x-axis so it has no x variable. In the yz-plane it is the line through $(0, 0, 4)$ and $(0, 2, 0)$: $z + 2y = 4$. The plane is $z + 2y = 4$.

5) Find the equation of the sphere with center $(2, 3, -4)$ and radius 5.

$(x - 2)^2 + (y - 3)^2 + (z + 4)^2 = 25.$

6) Find the center and radius of the sphere $x^2 - 6x + y^2 + 4y + z^2 = 1$.

Complete the square for x and y:
$x^2 - 6x + 9 + y^2 + 4y + 4 + z^2$
$\qquad = 1 + 9 + 4$
$(x - 3)^2 + (y + 2)^2 + z^2 = 14$
center: $(3, -2, 0)$
radius: $\sqrt{14}$.

7) Describe the region given by each:

a) $x^2 + z^2 = 4$

This is a cylinder of radius 2 along the y-axis.

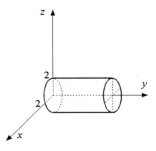

b) $x^2 + y^2 + z^2 \leq 1$

A solid ball of radius 1 and center $(0, 0, 0)$.

c) $\{(x, y, z): x = 3 \text{ and } y = 4\}$

This is a line through $(3, 4, 0)$ and parallel to the z-axis.

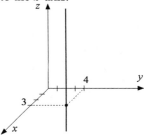

Vectors

Concepts to Master

A. Vectors; Position vector; Length; Unit vector; Standard basis vectors $(\mathbf{i}, \mathbf{j}, \mathbf{k})$
B. Vector arithmetic; Scalars; Vector properties

Summary and Focus Questions

A. A <u>vector</u> is an object having both a length and a direction.

A <u>two-dimensional vector</u> is an ordered pair $\mathbf{a} = \langle a_1, a_2 \rangle$. A <u>three-dimensional vector</u> is an ordered triple $\mathbf{a} = \langle a_1, a_2, a_3 \rangle$.

The numbers a_i are called components of $\mathbf{a}$. The directed line segment $\overrightarrow{AB}$ from A: (x_1, y_1) to B: (x_2, y_2) represents the vector $\mathbf{a} = \langle x_2 - x_1, y_2 - y_1 \rangle$.

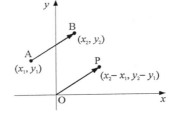

$\mathbf{a}$ is the <u>position vector</u> for the line segment from the origin O to P: $(x_2 - x_1, y_2 - y_1)$.

For $\mathbf{a} = \langle a_1, a_2 \rangle$, the <u>length of $\mathbf{a}$</u> is $|\mathbf{a}| = \sqrt{a_1^2 + a_2^2}$. For $\mathbf{a} = \langle a_1, a_2, a_3 \rangle$ the <u>length of $\mathbf{a}$</u> is $|\mathbf{a}| = \sqrt{a_1^2 + a_2^2 + a_3^2}$.
A <u>unit vector</u> has length 1. The <u>zero vector</u> is $\mathbf{0} = \langle 0, 0 \rangle$ or $\mathbf{0} = \langle 0, 0, 0 \rangle$.

In two dimensions $\mathbf{i} = \langle 1, 0 \rangle$ and $\mathbf{j} = \langle 0, 1 \rangle$ are the <u>standard basis vectors</u>. For $\mathbf{a} = \langle a_1, a_2 \rangle$, we can write $\mathbf{a} = a_1\mathbf{i} + a_2\mathbf{j}$. (We will have more to say about vector addition in part B.)

49

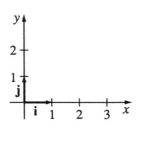

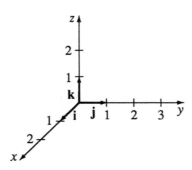

In three-dimensional space $\mathbf{i} = \langle 1, 0, 0 \rangle$, $\mathbf{j} = \langle 0, 1, 0 \rangle$, $\mathbf{k} = \langle 0, 0, 1 \rangle$ and any $\mathbf{a} = \langle a_1, a_2, a_3 \rangle$ may be written $\mathbf{a} = a_1\mathbf{i} + a_2\mathbf{j} + a_3\mathbf{k}$.

1) The vector represented by the line segment from $(5, 6)$ to $(3, 7)$ is

    _____.

$$\langle 3 - 5, 7 - 6 \rangle = \langle -2, 1 \rangle.$$

2) The vector represented by the directed line segment from $(10, 4, 3)$ to $(5, 4, 6)$ is _____.

$$\langle 5 - 10, 4 - 4, 6 - 3 \rangle = \langle -5, 0, 3 \rangle.$$

3) Sketch the position vectors for each:

    a) $\mathbf{a} = \langle 2, -6 \rangle$.

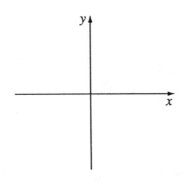

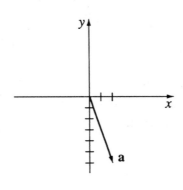

b) $\mathbf{a} = \langle 3,\ 6,\ 4 \rangle$.

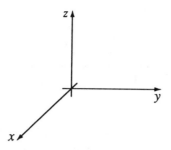

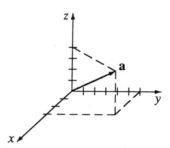

4)   The length of $\mathbf{a} = \langle 4,\ 2,\ -2 \rangle$ is ____.

$$|\mathbf{a}| = \sqrt{4^2 + 2^2 + (-2)^2} = \sqrt{24}.$$

5)   Is $\mathbf{a} = \left\langle \frac{-1}{4},\ \frac{\sqrt{3}}{2},\ \frac{\sqrt{3}}{4} \right\rangle$ a unit vector?

Yes, since

$$|\mathbf{a}| = \sqrt{\left(\frac{-1}{4}\right)^2 + \left(\frac{\sqrt{3}}{2}\right)^2 + \left(\frac{\sqrt{3}}{4}\right)^2} = 1.$$

6)   Write each in terms of unit basis vectors:

   a) $\mathbf{a} = \langle 4,\ -3 \rangle$.

$$\mathbf{a} = 4\mathbf{i} - 3\mathbf{j}$$

   b) $\mathbf{a} = \langle 0,\ 4,\ 5 \rangle$.

$$\mathbf{a} = 4\mathbf{j} + 5\mathbf{k}$$

   c) $\mathbf{a}$ is represented by $\overrightarrow{AB}$ where
        A: $(2, 5)$, B: $(6, 1)$.

$$\mathbf{a} = \langle 6 - 2,\ 1 - 5 \rangle = \langle 4,\ -4 \rangle = 4\mathbf{i} - 4\mathbf{j}$$

B. Vector addition and scalar multiplication are done componentwise:
   For $\mathbf{a} = \langle a_1,\ a_2 \rangle$ and $\mathbf{b} = \langle b_1,\ b_2 \rangle$ and real number c (called a <u>scalar</u>),
   $$\mathbf{a} + \mathbf{b} = \langle a_1 + b_1,\ a_2 + b_2 \rangle.$$
   $$c\mathbf{a} = \langle ca_1,\ ca_2 \rangle$$

a + **b** is the vector obtained by
positioning the initial point of **b** at
the terminal point of **a** and drawing
the directed line segment from the
initial point of **a** to the terminal
point of **b**.

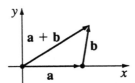

For $c > 0$, c**a** is the vector in the same direction as **a** with length c times the
length of **a**. For $c < 0$, c**a** is the vector in the opposite direction of **a** with
length $-c$ times the length of **a**.

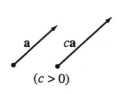

a − **b** is defined to be
a + (−**b**) = **a** + (−1)**b**.

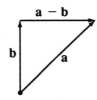

For any nonzero vector **a**, $\frac{\mathbf{a}}{|\mathbf{a}|}$ is a unit vector. Nonzero vector **a** is parallel to **b**
means **b** = c**a** for some c.

Similar definitions and interpretations hold for vectors in three dimensions.

These definitions allow us to write **a** = $\langle a_1, a_2 \rangle$ as **a** = $a_1\mathbf{i} + a_2\mathbf{j}$ because
$\langle a_1, a_2 \rangle = \langle a_1, 0 \rangle + \langle 0, a_2 \rangle = a_1\langle 1, 0 \rangle + a_2\langle 0, 1 \rangle = a_1\mathbf{i} + a_2\mathbf{j}$.

Properties of vectors:

$$\mathbf{a} + \mathbf{b} = \mathbf{b} + \mathbf{a}$$
$$\mathbf{a} + (\mathbf{b} + \mathbf{c}) = (\mathbf{a} + \mathbf{b}) + \mathbf{c}$$
$$\mathbf{a} + \mathbf{0} = \mathbf{a} \quad (\mathbf{0} \text{ is the zero vector.})$$

$$c(\mathbf{a} + \mathbf{b}) = c\mathbf{a} + c\mathbf{b}$$
$$(c + d)\mathbf{a} = c\mathbf{a} + d\mathbf{a}$$
$$(cd)\mathbf{a} = c(d\mathbf{a})$$

Vectors and their operations in higher dimensions are defined similarly. For example, in 4 dimensions with the 4 unit basis vectors $\mathbf{u}_1, \mathbf{u}_2, \mathbf{u}_3, \mathbf{u}_4$, a vector $\mathbf{a} = \langle a_1, a_2, a_3, a_4 \rangle$ may be written $\mathbf{a} = a_1\mathbf{u}_1 + a_2\mathbf{u}_2 + a_3\mathbf{u}_3 + a_4\mathbf{u}_4$ and $|\mathbf{a}| = \sqrt{a_1^2 + a_2^2 + a_3^2 + a_4^2}$. The set of all n-dimensional vectors is denoted V_n.

7) For $\mathbf{a} = \langle 4, 3 \rangle$ and $\mathbf{b} = \langle -5, 1 \rangle$:

   a) $\mathbf{a} + \mathbf{b} = $ _____.

            $\langle 4, 3 \rangle + \langle -5, 1 \rangle = \langle -1, 4 \rangle.$

   b) $6\mathbf{a} = $ _____.

            $\langle 6(4), 6(3) \rangle = \langle 24, 18 \rangle.$

   c)   Sketch the position vector of $\mathbf{b} - \mathbf{a}$.

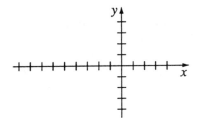

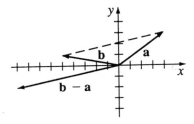

$$\mathbf{b} - \mathbf{a} = \langle -5, 1 \rangle - \langle 4, 3 \rangle = \langle -9, -2 \rangle.$$

8) True or False:
$$c(\mathbf{a} - \mathbf{b}) = c\mathbf{a} - c\mathbf{b}.$$

            True.

9) Find the unit vector in the direction of $\mathbf{a} = \langle 4, 3, -1 \rangle$.

$$|\mathbf{a}| = \sqrt{4^2 + 3^2 + (-1)^2} = \sqrt{26}.$$
The unit vector is
$$\frac{\mathbf{a}}{|\mathbf{a}|} = \left\langle \frac{4}{\sqrt{26}}, \frac{3}{\sqrt{26}}, \frac{-1}{\sqrt{26}} \right\rangle.$$

10) For $\mathbf{a} = 7\mathbf{i} + 4\mathbf{j} - 3\mathbf{k}$ and
$\mathbf{b} = 2\mathbf{i} + \mathbf{j} + 5\mathbf{k}$:

a) $\mathbf{a} + \mathbf{b} = $ _____.

b) $-2\mathbf{b} = $ _____.

11) Is $\mathbf{a} = \langle 4, -8, 6 \rangle$ parallel to
$\mathbf{b} = \langle -2, 3, -4 \rangle$?

$9\mathbf{i} + 5\mathbf{j} + 2\mathbf{k}$

$-4\mathbf{i} - 2\mathbf{j} - 10\mathbf{k}$

No. Comparing first components $\mathbf{a} = -2\mathbf{b}$ if
$\mathbf{a}$ and $\mathbf{b}$ are parallel. But comparing second
components, $\mathbf{a} = -\frac{8}{3}\mathbf{b}$.

12) A plane is heading north at 250 mph but
there is a 40 mph cross wind from the
west. What is the true direction the
plane is heading?

Let $\mathbf{u}$ be the vector representing due north at
250 mph. Let $\mathbf{v}$ be the wind from the west.
The true heading is
$\mathbf{u} + \mathbf{v} = \langle 0, 250 \rangle + \langle 40, 0 \rangle = \langle 40, 250 \rangle$.

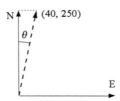

For the angle θ from north
$\tan \theta = \frac{40}{250} = 0.16$

$\theta = \tan^{-1} 0.16 = 0.159$
The true course is 0.159 radians ($\approx 9°$) east
of due north.

The Dot Product

Concepts to Master

A. Dot product; properties of $\mathbf{a} \cdot \mathbf{b}$

B. Angle between two vectors; Projection; Work; Direction cosines

Summary and Focus Questions

A. For $\mathbf{a} = \langle a_1, a_2 \rangle$ and $\mathbf{b} = \langle b_1, b_2 \rangle$, the <u>dot product</u> is $\mathbf{a} \cdot \mathbf{b} = a_1 b_1 + a_2 b_2$. For $\mathbf{a} = \langle a_1, a_2, a_3 \rangle$ and $\mathbf{b} = \langle b_1, b_2, b_3 \rangle$, the <u>dot product</u> $\mathbf{a} \cdot \mathbf{b}$ is $\mathbf{a} \cdot \mathbf{b} = a_1 b_1 + a_2 b_2 + a_3 b_3$.

Properties of the dot product include

$\mathbf{a} \cdot \mathbf{b} = \mathbf{b} \cdot \mathbf{a}$

$\mathbf{a} \cdot (\mathbf{b} + \mathbf{c}) = \mathbf{a} \cdot \mathbf{b} + \mathbf{a} \cdot \mathbf{c}$

$c(\mathbf{a} \cdot \mathbf{b}) = (c\mathbf{a}) \cdot \mathbf{b}$

$\mathbf{a} \cdot \mathbf{a} = |\mathbf{a}|^2$

1) Find $\mathbf{a} \cdot \mathbf{b}$ for

   a) $\mathbf{a} = \langle 6, 3 \rangle$, $\mathbf{b} = \langle 2, -1 \rangle$

   $\mathbf{a} \cdot \mathbf{b} = 6(2) + 3(-1) = 9.$

   b) $\mathbf{a} = 4\mathbf{i} - 3\mathbf{j} + \mathbf{k}$, $\mathbf{b} = 5\mathbf{j} + 10\mathbf{k}$

   $\mathbf{a} \cdot \mathbf{b} = 4(0) + (-3)5 + 1(10) = -5.$

   c) $\mathbf{a} = \langle 8, 1, 4 \rangle$, $\mathbf{b} = \langle 3, 0, -6 \rangle$.

   $\mathbf{a} \cdot \mathbf{b} = 8(3) + 1(0) + 4(-6) = 0.$

2) Why is $\mathbf{a} \cdot \mathbf{b} \cdot \mathbf{c}$ not defined although $(\mathbf{a} \cdot \mathbf{b})\mathbf{c}$ is defined?

   $\mathbf{a} \cdot \mathbf{b} \cdot \mathbf{c}$ makes no sense because the dot product $\mathbf{a} \cdot \mathbf{b}$ is a scalar and you can not compute the dot product of a scalar with a vector $\mathbf{c}$. But $(\mathbf{a} \cdot \mathbf{b})\mathbf{c}$ [with no dot between the) and $\mathbf{c}$] is a scalar multiple of $\mathbf{c}$.

B. The <u>angle θ between vectors **a** and **b**</u> satisfies
$\mathbf{a} \cdot \mathbf{b} = |\mathbf{a}|\,|\mathbf{b}|\cos\theta$, or equivalently, $\cos$
$\theta = \frac{\mathbf{a}\cdot\mathbf{b}}{|\mathbf{a}|\,|\mathbf{b}|}$.

a is <u>orthogonal</u> (<u>perpendicular</u>) to **b** iff $\mathbf{a} \cdot \mathbf{b} = 0$.

The <u>scalar projection</u> of **b** on **a** is the number $|\mathbf{b}|\cos\theta$. Intuitively, the scalar projection is the length of the shadow of **b** cast upon **a** by a light source directly over **a**.

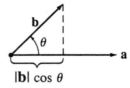

The <u>work</u> done by a constant force **F** moving an object along a directed line segment represented by a vector **a** is $\mathbf{F} \cdot \mathbf{a}$.

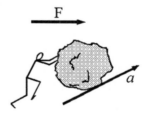

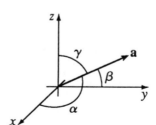

For nonzero $\mathbf{a} = (a_1,\ a_2,\ a_3)$, let
    α = angle between **a** and positive x-axis,
    β = angle between **a** and positive y-axis,
    γ = angle between **a** and positive z-axis.

$\cos\alpha = \frac{a_1}{|\mathbf{a}|}$, $\cos\beta = \frac{a_2}{|\mathbf{a}|}$, $\cos\gamma = \frac{a_3}{|\mathbf{a}|}$ are the
<u>direction cosines</u> for the direction angles α, β, and γ.

3) Find the angle θ between the vectors
$\mathbf{a} = \langle 1, -1, -1 \rangle$ and $\mathbf{b} = \langle -1, 5, 1 \rangle$.

$$\cos \theta = \frac{\mathbf{a} \cdot \mathbf{b}}{|\mathbf{a}|\,|\mathbf{b}|}$$
$$= \frac{1(-1)+(-1)5+(-1)1}{\sqrt{1^2+(-1)^2+(-1)^2}\sqrt{(-1)^2+5^2+1^2}}$$
$$= \frac{-7}{9}.$$
Therefore,
$\theta = \cos^{-1}\left(\frac{-7}{9}\right) \approx 2.46$ rad $\approx 141°$.

4) Find the scalar projection of
$\mathbf{x} = 3\mathbf{i} + \mathbf{j} + \mathbf{k}$ along
$\mathbf{y} = 2\mathbf{i} + 3\mathbf{j} - 4\mathbf{k}$.

$|\mathbf{x}| = \sqrt{11}$, $|\mathbf{y}| = \sqrt{29}$. For θ, the angle
between $\mathbf{x}$ and $\mathbf{y}$, $\cos \theta = \frac{\mathbf{x} \cdot \mathbf{y}}{|\mathbf{x}|\,|\mathbf{y}|} = \frac{5}{\sqrt{11}\sqrt{29}}$.
The projection is
$|\mathbf{x}| \cos \theta = \sqrt{11}\frac{5}{\sqrt{11}\sqrt{29}} = \frac{5}{\sqrt{29}}$.

5) Is $6\mathbf{i} - 3\mathbf{j} + 2\mathbf{k}$ orthogonal to
$2\mathbf{i} + 10\mathbf{j} + 9\mathbf{k}$?

Yes. the dot product is
$6(2) + (-3)10 + 2(9) = 0$.

6) Find the amount of work done by a
horizontal force of 4 newtons moving
an object 3 meters up a hill that makes
a 30° angle with the horizontal ground.
(Disregard gravity.)

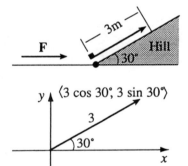

The vector representing the distance
travelled is
$\mathbf{d} = \langle 3 \cos 30°, 3 \sin 30° \rangle = \left\langle \frac{3\sqrt{3}}{2}, \frac{3}{2} \right\rangle$.
The force is $\mathbf{F} = \langle 4, 0 \rangle$ so the work done is
$\mathbf{F} \cdot \mathbf{d} = 4\left(\frac{3\sqrt{3}}{2}\right) + 0\left(\frac{3}{2}\right) = 6\sqrt{3}$ joules.

7) Find the direction cosines of
$\mathbf{a} = 3\mathbf{i} - \mathbf{j} + 4\mathbf{k}$.

$|\mathbf{a}| = \sqrt{3^2 + (-1)^2 + 4^2} = \sqrt{26}$.
$\cos \alpha = \frac{3}{\sqrt{26}}$, $\cos \beta = \frac{-1}{\sqrt{26}}$, $\cos \gamma = \frac{4}{\sqrt{26}}$.

8) Find any nonzero vector **x** orthogonal to **y** = $\langle 3, 4 \rangle$

If **x** = $\langle a, b \rangle$ is orthogonal to y, then **x** · **y** = $3a + 4b = 0$. Any solution other than $(0, 0)$ will do. Let $a = 4$ and $b = -3$. Then **x** = $\langle 4, -3 \rangle$.

9) Find a unit vector orthogonal to **a** = $\langle 4, -3, 6 \rangle$.

First, find a vector **b** = $\langle x, y, z \rangle$ where **a** · **b** = 0. **a** · **b** = $4x - 3y + 6z = 0$. One (of infinitely many) solution is $x = 3$, $y = 4$, $z = 0$, thus **b** = $\langle 4, 3, 0 \rangle$. $|\mathbf{b}| = \sqrt{4^2 + 3^2 + 0^2} = 5$. The unit vector is $\frac{\mathbf{b}}{|\mathbf{b}|} = \left\langle \frac{4}{5}, \frac{3}{5}, 0 \right\rangle$.

The Cross Product

Concepts to Master

A. Cross product; Properties of $\mathbf{a} \times \mathbf{b}$

B. Area of parallelogram; Scalar triple product; Volume of parallelepiped

---

Summary and Focus Questions

A. The <u>cross product $\mathbf{a} \times \mathbf{b}$</u> of vectors $\mathbf{a} = \langle a_1, a_2, a_3 \rangle$ and $\mathbf{b} = \langle b_1, b_2, b_3 \rangle$ is
$\mathbf{a} \times \mathbf{b} = \langle a_2 b_3 - a_3 b_2, \ a_3 b_1 - a_1 b_3, \ a_1 b_2 - a_2 b_1 \rangle$.

Cross product is defined only for three-dimensional vectors. Computing $\mathbf{a} \times \mathbf{b}$ is easier if written in terms of determinants. For a 2 by 2 array (a square of four numbers) the determinant of order 2 is:
$$\begin{vmatrix} x & y \\ z & w \end{vmatrix} = xw - zy$$
which is often thought of as the product of the numbers on the upper left to lower right "diagonal" (xw) minus the product from the lower left to upper right "diagonal" (zy).

For example $\begin{vmatrix} 6 & 2 \\ 3 & 5 \end{vmatrix} = 6(5) - 3(2) = 30 - 6 = 24$.

A determinant of order 3 (from a square array of 9 numbers) may be used as a handy way to compute $\mathbf{a} \times \mathbf{b}$, where we use vectors $\mathbf{i}, \mathbf{j}, \mathbf{k}$ as symbols in the first row of the determinant, the components of $\mathbf{a}$ in the second row and the components of $\mathbf{b}$ in the third row.

For example, if $\mathbf{a} = \langle 2, 1, 5 \rangle$ and $\mathbf{b} = \langle 4, 6, 3 \rangle$ then

$$\mathbf{a} \times \mathbf{b} = \begin{vmatrix} \mathbf{i} & \mathbf{j} & \mathbf{k} \\ 2 & 1 & 5 \\ 4 & 6 & 3 \end{vmatrix}.$$

This is evaluated as follows:

the determinant obtained by deleting the row and column of **i** times **i**

$$\begin{vmatrix} 1 & 5 \\ 6 & 3 \end{vmatrix} \mathbf{i} = (1 \cdot 3 - 6 \cdot 5)\mathbf{i} = -27\mathbf{i}$$

minus

the determinant obtained by deleting the row and column of **j** times **j**

$$\begin{vmatrix} 2 & 5 \\ 4 & 3 \end{vmatrix} \mathbf{j} = (2 \cdot 3 - 4 \cdot 5)\mathbf{j} = -14\mathbf{j}$$

plus

the determinant obtained by deleting the row and column of **k** times **k**.

$$+\begin{vmatrix} 2 & 1 \\ 4 & 6 \end{vmatrix} \mathbf{k} = (2 \cdot 6 - 4 \cdot 1)\mathbf{k} = 8\mathbf{k}.$$

Thus $\mathbf{a} \times \mathbf{b} = -27\mathbf{i} - (-14\mathbf{j}) + 8\mathbf{k} = -27\mathbf{i} + 14\mathbf{j} + 8\mathbf{k}.$

Properties of cross products:

$\mathbf{a} \times \mathbf{b}$ is orthogonal to both **a** and **b**.

$|\mathbf{a} \times \mathbf{b}| = |\mathbf{a}|\,|\mathbf{b}| \sin \theta$, θ = the angle between **a** and **b**.

$\mathbf{a} \times \mathbf{b} = -\mathbf{b} \times \mathbf{a}$

$c(\mathbf{a} \times \mathbf{b}) = c\mathbf{a} \times \mathbf{b} = \mathbf{a} \times c\mathbf{b}$ $\qquad$ (c, a scalar)

$\mathbf{a} \times (\mathbf{b} + \mathbf{c}) = \mathbf{a} \times \mathbf{b} + \mathbf{a} \times \mathbf{c}$ $\;$ and $\;$ $(\mathbf{a} + \mathbf{b}) \times \mathbf{c} = \mathbf{a} \times \mathbf{c} + \mathbf{b} \times \mathbf{c}.$

$\mathbf{a} \times (\mathbf{b} \times \mathbf{c}) = (\mathbf{a} \cdot \mathbf{c})\mathbf{b} - (\mathbf{a} \cdot \mathbf{b})\mathbf{c}$

$\mathbf{i} \times \mathbf{j} = \mathbf{k},$ $\;$ $\mathbf{j} \times \mathbf{k} = \mathbf{i},$ $\;$ and $\mathbf{k} \times \mathbf{i} = \mathbf{j}$

1) Find $\mathbf{a} \times \mathbf{b}$, where $\mathbf{a} = \langle 4, -1, 3 \rangle$ and $\mathbf{b} = \langle 1, 6, 2 \rangle$.

$$\mathbf{a} \times \mathbf{b} = \begin{vmatrix} \mathbf{i} & \mathbf{j} & \mathbf{k} \\ 4 & -1 & 3 \\ 1 & 6 & 2 \end{vmatrix}$$

$$= \begin{vmatrix} -1 & 3 \\ 6 & 2 \end{vmatrix} \mathbf{i} - \begin{vmatrix} 4 & 3 \\ 1 & 2 \end{vmatrix} \mathbf{j} + \begin{vmatrix} 4 & -1 \\ 1 & 6 \end{vmatrix} \mathbf{k}$$

$$= -20\mathbf{i} - 5\mathbf{j} + 25\mathbf{k}.$$

2) For any **a**, **b** what is $\mathbf{a} \times \mathbf{b} + \mathbf{b} \times \mathbf{a}$?

Since $\mathbf{a} \times \mathbf{b} = -\mathbf{b} \times \mathbf{a}$, $\mathbf{a} \times \mathbf{b} + \mathbf{b} \times \mathbf{a} = \mathbf{0}$, the zero vector.

3) Evaluate $2\mathbf{i} \times (4\mathbf{j} + 3\mathbf{k})$.

This may be done using determinants but we use cross product properties:
$$2\mathbf{i} \times (4\mathbf{j} + 3\mathbf{k}) = 2\mathbf{i} \times 4\mathbf{j} + 2\mathbf{i} \times 3\mathbf{k}$$
$$= 8(\mathbf{i} \times \mathbf{j}) + 6(\mathbf{i} \times \mathbf{k})$$
$$= 8(\mathbf{i} \times \mathbf{j}) - 6(\mathbf{k} \times \mathbf{i})$$
$$= 8\mathbf{k} - 6\mathbf{j}.$$

4) Find a vector orthogonal to both $2\mathbf{i} - 3\mathbf{j} + \mathbf{k}$ and $\mathbf{i} + \mathbf{j} + 2\mathbf{k}$.

The cross product will do
$$\begin{vmatrix} \mathbf{i} & \mathbf{j} & \mathbf{k} \\ 2 & -3 & 1 \\ 1 & 1 & 2 \end{vmatrix}$$
$$= \begin{vmatrix} -3 & 1 \\ 1 & 2 \end{vmatrix} \mathbf{i} - \begin{vmatrix} 2 & 1 \\ 1 & 2 \end{vmatrix} \mathbf{j} + \begin{vmatrix} 2 & -3 \\ 1 & 1 \end{vmatrix} \mathbf{k}$$
$$= -7\mathbf{i} - 3\mathbf{j} + 5\mathbf{k}.$$

B. $|\mathbf{a} \times \mathbf{b}|$ is the area of the parallelogram formed by $\mathbf{a}$ and $\mathbf{b}$.

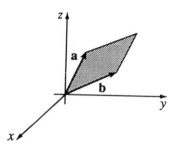

The triple scalar product of vectors $\mathbf{a} = \langle a_1, a_2, a_3 \rangle$, $\mathbf{b} = \langle b_1, b_2, b_3 \rangle$, $\mathbf{c} = \langle c_1, c_2, c_3 \rangle$ is the number $\mathbf{a} \cdot (\mathbf{b} \times \mathbf{c})$. It may be evaluated directly or by using
$$\mathbf{a} \cdot (\mathbf{b} \times \mathbf{c}) = \begin{vmatrix} a_1 & a_2 & a_3 \\ b_1 & b_2 & b_3 \\ c_1 & c_2 & c_3 \end{vmatrix}.$$

$|\mathbf{a} \cdot (\mathbf{b} \times \mathbf{c})|$ is the volume of the parallelepiped formed by $\mathbf{a}$, $\mathbf{b}$, and $\mathbf{c}$.

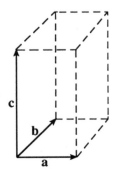

5) Find the area of the parallelogram with vertices $P:(4, 1, 3)$, $Q:(7, 5, 3)$, $R:(6, 4, 2)$, $S:(9, 8, 2)$.

Let $\mathbf{a} = \overrightarrow{PQ} = \langle 7 - 4, 5 - 1, 3 - 3 \rangle$
$\qquad = \langle 3, 4, 0 \rangle$
and $\mathbf{b} = \overrightarrow{PR} = \langle 6 - 4, 4 - 1, 2 - 3 \rangle$
$\qquad = \langle 2, 3, -1 \rangle$.
The area we seek is $|\mathbf{a} \times \mathbf{b}|$.

$$\mathbf{a} \times \mathbf{b} = \begin{vmatrix} \mathbf{i} & \mathbf{j} & \mathbf{k} \\ 3 & 4 & 0 \\ 2 & 3 & -1 \end{vmatrix} = -4\mathbf{i} + 3\mathbf{j} + \mathbf{k}.$$

Thus $|\mathbf{a} \times \mathbf{b}| = \sqrt{(-4)^2 + 3^2 + (-1)^2} = \sqrt{26}$.

6) Find $\mathbf{a} \cdot (\mathbf{b} \times \mathbf{c})$ where $\mathbf{a} = \langle 1, 3, 1 \rangle$, $\mathbf{b} = \langle 0, 4, 2 \rangle$, $\mathbf{c} = \langle -2, 2, 3 \rangle$:

a) directly

$$\mathbf{b} \times \mathbf{c} = \begin{vmatrix} \mathbf{i} & \mathbf{j} & \mathbf{k} \\ 0 & 4 & 2 \\ -2 & 2 & 3 \end{vmatrix}$$

$$= \begin{vmatrix} 4 & 2 \\ 2 & 3 \end{vmatrix} \mathbf{i} - \begin{vmatrix} 0 & 2 \\ -2 & 3 \end{vmatrix} \mathbf{j} + \begin{vmatrix} 0 & 4 \\ -2 & 2 \end{vmatrix} \mathbf{k}$$

$$= 8\mathbf{i} - 4\mathbf{j} + 8\mathbf{k}.$$

Thus $\mathbf{a} \cdot (\mathbf{b} \times \mathbf{c}) = 1(8) + 3(-4) + 1(8) = 4$.

b) using determinants

$$\mathbf{a} \cdot (\mathbf{b} \times \mathbf{c}) = \begin{vmatrix} 1 & 3 & 1 \\ 0 & 4 & 2 \\ -2 & 2 & 3 \end{vmatrix}$$

$$= 1\begin{vmatrix} 4 & 2 \\ 2 & 3 \end{vmatrix} - 3\begin{vmatrix} 0 & 2 \\ -2 & 3 \end{vmatrix} + 1\begin{vmatrix} 0 & 4 \\ -2 & 2 \end{vmatrix}$$

$$= 1(8) - 3(4) + 1(8) = 4.$$

7) Find the volume of the parallelipiped formed by $\mathbf{a} = -2\mathbf{i} + 3\mathbf{j}$, $\mathbf{b} = 4\mathbf{i} + \mathbf{j} - \mathbf{k}$, $\mathbf{c} = 6\mathbf{j} + \mathbf{k}$.

$$\mathbf{a} \cdot (\mathbf{b} \times \mathbf{c}) = \begin{vmatrix} -2 & 3 & 0 \\ 4 & 1 & -1 \\ 0 & 6 & 1 \end{vmatrix}$$

$$= -2\begin{vmatrix} 1 & -1 \\ 6 & 1 \end{vmatrix} - 3\begin{vmatrix} 4 & -1 \\ 0 & 1 \end{vmatrix} + 0\begin{vmatrix} 4 & 1 \\ 0 & 6 \end{vmatrix}$$

$$= -2(7) - 3(4) + 0 = -26$$

$|\mathbf{a} \cdot (\mathbf{b} \times \mathbf{c})| = 26$.

Equations of Lines and Planes

Concepts to Master

A. Equations of lines in space (vector, parametric, and symmetric); Parallel and skew lines

B. Normal vector; Equations of planes (vector and scalar); Parallel planes; Angle between intersecting planes; Distance from a point to a plane and distance between two planes

Summary and Focus Questions

A. Let L be a line passing through P_0: (x_0, y_0, z_0) and parallel to vector $\mathbf{v} = \langle a, b, c \rangle$. Let $\mathbf{r}_0 = \langle x_0, y_0, z_0 \rangle$. The three forms of the equations for L are

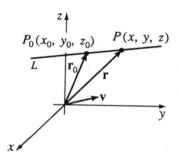

Vector Form	**Parametric Form**	**Symmetric Form**
$\mathbf{r} = \mathbf{r}_0 + t\mathbf{v}$	$x = x_0 + at$	$\dfrac{x - x_0}{a} = \dfrac{y - y_0}{b} = \dfrac{z - z_0}{c}$
	$y = y_0 + bt$	a, b, c nonzero
	$z = z_0 + ct$	

In case, for example, $b = 0$, the symmetric form is modified to $y = y_0$ and $\dfrac{x - x_0}{a} = \dfrac{z - z_0}{c}$.

For lines $L_1 : \mathbf{r} = \mathbf{r}_0 + t_1\mathbf{v}_1$ and $L_2 : \mathbf{r} = \mathbf{s}_0 + t_2\mathbf{v}_2$
   1) L_1 and L_2 are parallel iff $\mathbf{v}_1$ is a scalar multiple of $\mathbf{v}_2$.
   2) L_1 and L_2 intersect iff $\mathbf{r}_0 + t_1\mathbf{v}_1 = \mathbf{s}_0 + t_2\mathbf{v}_2$ for some t_1, t_2.
Lines that are not parallel and do not intersect are called <u>skew</u>.

1) Find the three forms for the equation of the line L through $(4, 1, 6)$ parallel to $\langle 7, 3, 0 \rangle$.

Vector form: $\mathbf{r} = \langle 4, 1, 6 \rangle + t \langle 7, 3, 0 \rangle$, or $\mathbf{r} = (4 + 7t)\mathbf{i} + (1 + 3t)\mathbf{j} + 6\mathbf{k}$.

Parametric form: $x = 4 + 7t$
$$y = 1 + 3t$$
$$z = 6 + 0t \quad (\text{or } z = 6)$$

Symmetric form: $z = 6$ and $\frac{x-4}{7} = \frac{y-1}{3}$.

2) Determine whether the lines L_1 and L_2 are parallel, intersect, or are skew:

L_1: $x = 1 + 5t$ L_2: $x = 10 + 3s$
$\quad\;\; y = 3 + 4t$ $\quad\quad\;\; y = 17 - s$
$\quad\;\; z = -2 + t$ $\quad\quad\;\; z = -3 + 2s$

The direction $\langle 5, 4, 1 \rangle$ is not a multiple of the direction $\langle 3, -1, 2 \rangle$ so L_1 and L_2 are not parallel.
If L_1 and L_2 intersect, then
$$1 + 5t = 10 + 3s$$
$$3 + 4t = 17 - s$$
$$-2 + t = -3 + 2s$$
must have a solution. Multiplying the second equation by 2 and adding to the third gives
$$4 + 9t = 31,$$
$$9t = 27, \; t = 3.$$
Using $t = 3$ in the third,
$$-2 + 3 = -3 + 2s.$$
Thus $2s = 4$, $s = 2$.
The solution to the second and third equations is $t = 3$, $s = 2$. Since this also satisfies the first $[1 + 5(3) = 10 + 3(2)]$ the lines intersect. At $t = 3$, $x = 16$, $y = 15$, $z = 1$ so the point of intersection is $(16, 15, 1)$.

3) Find the equation of the line parallel to the given line through the point P. Use the same form as the equation for L.

a) $L: \frac{x-1}{8} = \frac{y+2}{5} = \frac{z+3}{-2}$
   $P: (4, 5, 1)$

   $\frac{x-4}{8} = \frac{y-5}{5} = \frac{z-1}{-2}.$

b) $L: \mathbf{r} = (3 + 4t)\mathbf{i} - t\mathbf{j} + (8 + t)\mathbf{k}$
   $P: (-1, 3, 2)$

   $\mathbf{r} = (-1 + 4t)\mathbf{i} + (3 - t)\mathbf{j} + (2 + t)\mathbf{k}.$

c) $L: \quad x = 9 - 9t$
   $\qquad y = 8 + 3t$
   $\qquad z = 2 + 4t$
   $P: (7, -3, 5)$

   $x = 7 - 9t.$
   $y = -3 + 3t.$
   $z = 5 + 4t.$

B. A vector orthogonal to a plane is called a <u>normal vector</u>.

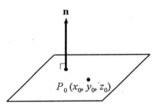

Suppose a plane contains a point $P: (x_0, y_0, z_0)$ and has a normal vector $\mathbf{n} = \langle a, b, c \rangle$.

Let $\mathbf{r}_0 = \langle x_0, y_0, z_0 \rangle$. The two forms of the equation of the plane are

Vector Form	**Scalar Form**
$\mathbf{n r} = \mathbf{n r}_0$	$a(x - x_0) + b(y - y_0) + c(z - z_0) = 0$

Every linear equation of the form $Ax + By + Cz = D$ is the equation of a plane in three-dimensional space.

If $\mathbf{a}$ and $\mathbf{b}$ are nonparallel vectors in a plane then $\mathbf{a} \times \mathbf{b}$ may be used as a normal vector for the plane.

Two planes are parallel if their normal vectors are parallel.

The angle θ between two intersecting planes with normal vectors $\mathbf{n}_1$ and $\mathbf{n}_2$ satisfies
$$\cos\theta = \frac{\mathbf{n}_1\cdot\mathbf{n}_2}{|\mathbf{n}_1|\,|\mathbf{n}_2|}$$

The distance from a point P: $(x_0,\,y_0,\,z_0)$ to a plane $ax+by+cz=d$ is
$$\frac{|ax_0+by_0+cz_0-d|}{\sqrt{a^2+b^2+c^2}}.$$

This formula may also be used to find the distance between two parallel planes, just let P be a point on one of the planes.

4) Find the two forms of the equation of the plane through $(8,\,2,\,3)$ with normal vector $\langle 6,\,2,\,5\rangle$.

Vector form:
$\langle 6,\,2,\,5\rangle\cdot\mathbf{r}=\langle 6,\,2,\,5\rangle\cdot\langle 8,\,2,\,3\rangle$
$\langle 6,\,2,\,5\rangle\cdot\mathbf{r}=67$

Scalar form:
$6(x-8)+2(y-2)+5(z-3)=0$
$6x+2y+5z=67$

5) Find the scalar equation of the plane through P: $(1,\,3,\,1)$, Q: $(3,\,0,\,4)$, R: $(4,\,-1,\,2)$.

Let $\mathbf{a}=\overrightarrow{PQ}=\langle 2,\,-3,\,3\rangle$ and $\mathbf{b}=\overrightarrow{PR}=\langle 3,\,-4,\,1\rangle$.
A normal to the plane is
$$\mathbf{a}\times\mathbf{b}=\begin{vmatrix} \mathbf{i} & \mathbf{j} & \mathbf{k}\\ 2 & -3 & 3\\ 3 & -4 & 1\end{vmatrix}=9\mathbf{i}+7\mathbf{j}+\mathbf{k}.$$
Using P as a point on the plane the equation is
$9(x-1)+7(y-3)+(z-1)=0$ or
$9x+7y+z=31.$

6) Find the cosine of the angle of intersection of the planes:
$3x+y+3z=5$
$6x-3y-2z=14$

Let $\mathbf{n}_1=\langle 3,\,1,\,3\rangle$, $\mathbf{n}_2=\langle 6,\,-3,\,-2\rangle$.
$\cos\theta=\frac{\mathbf{n}_1\cdot\mathbf{n}_2}{|\mathbf{n}_1|\,|\mathbf{n}_2|}=\frac{9}{\sqrt{19}\sqrt{49}}=\frac{9}{7\sqrt{19}}.$

7) Find the line of intersection of the planes: $2x + y - z = -4$
$3x + y + 2z = 5$

Let $\mathbf{n_1} = \langle 2, 1, -1 \rangle$, $\mathbf{n_2} = \langle 3, 1, 2 \rangle$. The line of intersection has the same direction as

$$\mathbf{n_1} \times \mathbf{n_2} = \begin{vmatrix} \mathbf{i} & \mathbf{j} & \mathbf{k} \\ 2 & 1 & -1 \\ 3 & 1 & 2 \end{vmatrix} = \langle 3, -7, -1 \rangle.$$

To find a point of intersection set $x = 0$ ($y = 0$ or $z = 0$ also could be used):
$$y - z = -4$$
$$y + 2z = 5$$
Subtracting: $3z = 9$, $z = 3$.
Thus $y - 3 = -4$, $y = -1$.
The line of intersection is $\frac{x}{3} = \frac{y+1}{-7} = \frac{z-3}{-1}$.

8) Find the distance from $(1, 2, -2)$ to $x + 3y + 4z = 12$.

$$\frac{|1(1)+3(2)+4(-2)-12|}{\sqrt{1^2+3^2+4^2}} = \frac{|-13|}{\sqrt{26}} = \frac{\sqrt{26}}{2}.$$

9) Find the distance between the parallel planes:
$2x + 3y + 7z = 10$ and
$2x + 3y + 7z = 134$.

$(5, 0, 0)$ is a point on the first plane. The distance between the planes is

$$\frac{|2(5)+3(0)+7(0)-134|}{\sqrt{2^2+3^2+7^2}} = \frac{|-124|}{\sqrt{62}}$$
$$= \frac{124}{\sqrt{62}} = 2\sqrt{62}.$$

Quadric Surfaces

Concepts to Master

Nine nondegenerate quadric surfaces

Summary and Focus Questions

The analogues of the three conics in two dimensions are nine <u>quadric surfaces</u> in three dimensions. Help in identifying a surface can come from examining a surfaces <u>traces</u> - its intersections with a plane parallel to one of the xy-, xz-, or yz-planes.

The table below and on the next pages summarizes the equations and their traces. Interchanging x, y, and z will produce surfaces of the same type but oriented along a different axis. For example $z^2 = \frac{x^2}{a^2} + \frac{y^2}{b^2}$, $y^2 = \frac{x^2}{a^2} + \frac{z^2}{c^2}$, and $x^2 = \frac{y^2}{b^2} + \frac{z^2}{c^2}$ are all elliptic cones.

Some equations may need to be rewritten by completing the square before trying to identify the surface.

Surface	Equation	Traces			Typical graph
		$xy \, (z = 0)$	$xz \, (y = 0)$	$yz \, (x = 0)$	
Ellipsoid	$\frac{x^2}{a^2} + \frac{y^2}{b^2} + \frac{z^2}{c^2} = 1$	ellipse	ellipse	ellipse	
Hyperboloid of One Sheet	$\frac{x^2}{a^2} + \frac{y^2}{b^2} - \frac{z^2}{c^2} = 1$	ellipse	hyperbola	hyperbola	

Surface	Equation	Traces			Typical graph
		$xy\,(z=0)$	$xz\,(y=0)$	$yz\,(x=0)$	
Hyperboloid of Two Sheets	$-\frac{x^2}{a^2} - \frac{y^2}{b^2} + \frac{z^2}{c^2} = 1$	none	hyperbola	hyperbola	
Elliptic Cone	$\frac{z^2}{c^2} = \frac{x^2}{a^2} + \frac{y^2}{b^2}$	$(0, 0, 0)$	two intersecting lines	two intersecting lines	
Elliptic Paraboloid	$\frac{z}{c} = \frac{x^2}{a^2} + \frac{y^2}{b^2}$	$(0, 0, 0)$	parabola upward	parabola upward	
Hyperbolic Paraboloid	$\frac{z}{c} = \frac{y^2}{b^2} - \frac{x^2}{a^2}$	two intersecting lines	parabola downward	parabola upward	

A cylinder will have one variable missing:

Surface	Equation	Traces			Typical graph
		$xy\,(z=0)$	$xz\,(y=0)$	$yz\,(x=0)$	
Parabolic Cylinder	$x^2 = 4ay$	parabola	z-axis	z-axis	

Surface	Equation	Traces			Typical graph
		$xy\,(z=0)$	$xz\,(y=0)$	$yz\,(x=0)$	
Elliptic Cylinder	$\frac{x^2}{a^2} + \frac{y^2}{b^2} = 1$	ellipse	two parallel lines	two parallel lines	
Hyperbolic Cylinder	$\frac{x^2}{a^2} - \frac{y^2}{b^2} = 1$	hyperbola	two parallel lines	none	

1) True or False:
   If one of x, y, z is missing from an equation, the surface is a cylinder.

   True.

2) True or False:
   A hyperbolic paraboloid must have an x^2 term, a y^2 term, and a z term.

   False. Two variables must be squared and the third one is not, but they need not be x^2, y^2, and z. For example, $y = \frac{x^2}{4} - \frac{z^2}{9}$ is a hyperbolic paraboloid.

3) Identify each:

   a) $\frac{x^2}{25} - \frac{y^2}{4} - \frac{z^2}{16} = 1$

   Hyperboloid of two sheets.

   b) $y - 4x^2 = 16z^2$

   Elliptic paraboloid.

   c) $z^2 + 10y = 0$

   Parabolic cylinder.

   d) $x^2 - 8x - y^2 - 2y + z^2 + 15 = 0$

   Complete the square:
   $x^2 - 8x + 16 - (y^2 + 2y + 1) + z^2 = 0$
   $(x - 4)^2 - (y + 1)^2 + z^2 = 0$
   $(y + 1)^2 = (x - 4)^2 + z^2$
   Elliptic cone.

e)

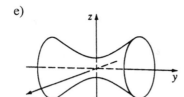

Hyperboloid of one sheet.

f)

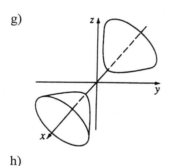

Hyperbolic paraboloid.

g)

Hyperboloid of two sheets.

h)

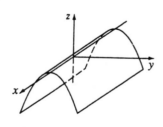

Parabolic cylinder.

4) Sometimes, Always, or Never:
    The linear terms are important in
    determining the type of quadric
    surface.

Sometimes. If the equation contains both a
first and a second power of a variable, the
presence of the first power does not affect
the type of surface it is.

Vector Functions and Space Curves

Concepts to Master

A. Vector functions; Limits; Parametric equations of a curve
B. Derivative of a vector function; Tangent vector; Properties of derivatives
C. Integrals of vector functions

Summary and Focus Questions

A. A <u>vector function</u> (or vector valued function) is a function having a set of real numbers as its domain and a set of vectors as its range; to each real number is associated a vector.

In three-dimensional space, a vector function may be written with component functions f, g, and h:
$$\mathbf{r}(t) = f(t)\mathbf{i} + g(t)\mathbf{j} + h(t)\mathbf{k}.$$

The graph of $\mathbf{r}(t)$ is a <u>curve</u> in space; we say the curve is defined parametrically by $x = f(t)$, $y = g(t)$, and $z = h(t)$.

$$\lim_{t \to t_0} \mathbf{r}(t) = (\lim_{t \to t_0} f(t))\mathbf{i} + (\lim_{t \to t_0} g(t))\mathbf{j} + (\lim_{t \to t_0} h(t))\mathbf{k}.$$
$\mathbf{r}(t)$ is <u>continuous at a</u> means $\lim_{t \to a} \mathbf{r}(t) = \mathbf{r}(a)$.

1) Sketch a graph of
$\mathbf{r}(t) = |\sin t|\mathbf{i} + |\sin t|\mathbf{j} + t\mathbf{k}.$

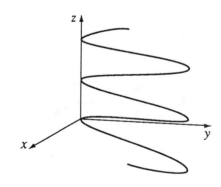

2) For $\mathbf{r}(t) = \langle \sin t,\ t^2 + 4,\ e^t \rangle$

  $\lim\limits_{t \to 0} \mathbf{r}(t) = $ _____ .

  $\langle \sin 0,\ 0^2 + 4,\ e^0 \rangle = \langle 0,\ 4,\ 1 \rangle.$

B. $\mathbf{r}'(t) = \lim\limits_{h \to 0} \dfrac{\mathbf{r}(t+h) - \mathbf{r}(t)}{h}$

  $= f'(t)\mathbf{i} + g'(t)\mathbf{j} + h'(k)\mathbf{k}.$

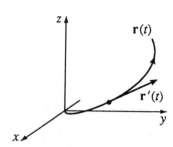

$\mathbf{r}'(t)$ is the tangent vector to the curve defined by $\mathbf{r}(t)$.

The unit tangent vector is

  $\mathbf{T}(t) = \dfrac{\mathbf{r}'(t)}{|\mathbf{r}'(t)|}.$

For differentiable vector functions $\mathbf{r}_1(t)$ and $\mathbf{r}_2(t)$ and scalar function $a(t)$:

  $[\mathbf{r}_1(t) + \mathbf{r}_2(t)]' = \mathbf{r}_1'(t) + \mathbf{r}_2'(t)$

  $[c\mathbf{r}_1(t)]' = c[\mathbf{r}_1'(t)] \quad (c,\ \text{a constant})$

  $[a(t)\mathbf{r}_1(t)]' = a'(t)\mathbf{r}_1(t) + a(t)\mathbf{r}_1'(t)$

  $[\mathbf{r}_1(t) \cdot \mathbf{r}_2(t)]' = \mathbf{r}_1'(t) \cdot \mathbf{r}_2(t) + \mathbf{r}_1(t) \cdot \mathbf{r}_2'(t)$

  $[\mathbf{r}_1(t) \times \mathbf{r}_2(t)]' = \mathbf{r}_1'(t) \times \mathbf{r}_2(t) + \mathbf{r}_1(t) \times \mathbf{r}_2'(t)$

  $[\mathbf{r}_1(a(t))]' = a'(t)\mathbf{r}_1'(a(t)) \qquad \text{(Chain Rule)}$

3) Find $\mathbf{r}'(t)$ for $\mathbf{r}(t) = \langle e^{2t},\ \tan t,\ t^3 \rangle$.

  $\mathbf{r}'(t) = \langle 2e^{2t},\ \sec^2 t,\ 3t^2 \rangle.$

4) Find the unit tangent vector to the curve $\mathbf{r}(t) = t^2\mathbf{i} - t^3\mathbf{j} + t^4\mathbf{k}$ at $t = 1$.

  $\mathbf{r}'(t) = 2t\mathbf{i} - 3t^2\mathbf{j} + 4t^3\mathbf{k}$
  $\mathbf{r}'(1) = 2\mathbf{i} - 3\mathbf{j} + 4\mathbf{k}$
  $\mathbf{T}(1) = \dfrac{2\mathbf{i} - 3\mathbf{j} + 4\mathbf{k}}{\sqrt{2^2 + (-3)^2 + 4^2}}$
  $\quad = \dfrac{2}{\sqrt{29}}\mathbf{i} - \dfrac{3}{\sqrt{29}}\mathbf{j} + \dfrac{4}{\sqrt{29}}\mathbf{k}.$

5) Find $\mathbf{r}''(t)$ for $\mathbf{r}(t) = \langle \cos t,\ t^3 \rangle$.

  $\mathbf{r}'(t) = \langle -\sin t,\ 3t^2 \rangle$
  $\mathbf{r}''(t) = \langle -\cos t,\ 6t \rangle$

6) Find $[\mathbf{f}(t) \cdot \mathbf{g}(t)]'$ for
$\mathbf{f}(t) = t^2\mathbf{i} + t^3\mathbf{j}$
$\mathbf{g}(t) = 7t\mathbf{i} + 3t^2\mathbf{j}$

$$\begin{aligned}
[\mathbf{f}(t) \cdot \mathbf{g}(t)]' &= \mathbf{f}'(t) \cdot \mathbf{g}(t) + \mathbf{f}(t) \cdot \mathbf{g}'(t) \\
&= [2t\mathbf{i} + 3t^2\mathbf{j}] \cdot [7t\mathbf{i} + 3t^2\mathbf{j}] \\
&\quad + [t^2\mathbf{i} + t^3\mathbf{j}] \cdot [7\mathbf{i} + 6t\mathbf{j}] \\
&= 14t^2 + 9t^4 + 7t^2 + 6t^4 \\
&= 21t^2 + 15t^4.
\end{aligned}$$

C. If $\mathbf{r}(t) = f(t)\mathbf{i} + g(t)\mathbf{j} + h(t)\mathbf{k}$ is continuous on $[a, \; b]$,

$$\int_a^b \mathbf{r}(t)\,dt = \left(\int_a^b f(t)\,dt\right)\mathbf{i} + \left(\int_a^b g(t)\,dt\right)\mathbf{j} + \left(\int_a^b h(t)\,dt\right)\mathbf{k},$$

or, written in vector form:
$$\int_a^b \mathbf{r}(t)\,dt = \mathbf{R}(b) - \mathbf{R}(a), \text{ where } \mathbf{R}'(t) = \mathbf{r}(t).$$

7) Evaluate $\int_1^3 \left(\mathbf{i} + 2t\mathbf{j} + \frac{1}{t}\mathbf{k}\right)dt.$

The integral equals
$$\left(\int_1^3 dt\right)\mathbf{i} + \left(\int_1^3 2t\,dt\right)\mathbf{j} + \left(\int_1^3 \frac{1}{t}\,dt\right)\mathbf{k}$$
$$= 2\mathbf{i} + 8\mathbf{j} + (\ln 3)\mathbf{k}$$

Arc Length and Curvature

Concepts to Master

A. Length of an arc in space; Smooth curve; Arc length function;
Reparametrization
B. Curvature; Unit Normal; Binormal

Summary and Focus Questions

A. A curve given by $\mathbf{r}(t)$ for $t \in [a, b]$ is <u>smooth</u> if $\mathbf{r}'$ is continuous and nonzero on (a, b). A curve made up of a finite number of smooth curves is called <u>piecewise smooth</u>.

If $\mathbf{r}(t) = \langle f(t), g(t), h(t) \rangle$ is smooth, the length of the curve is
$$L = \int_a^b |\mathbf{r}'(t)| dt = \int_a^b \sqrt{[f'(t)]^2 + [g'(t)]^2 + [h'(t)]^2} \; dt.$$

The arc length function is $s(t) = \int_a^t |\mathbf{r}'(u)| \, du$ and therefore $\frac{ds}{dt} = |\mathbf{r}'(t)|$.

A curve may be given by more than one function - these are different parametrizations.

If we can solve for t in terms of s (so that we can write $t = t(s)$), we can reparametrize the curve - that is, write it in the form $\mathbf{r} = \mathbf{r}(t(s))$.

1) Find the length of the arc given by
$\mathbf{r}(t) = \ln t\mathbf{i} - t^2\mathbf{j} + 2t\mathbf{k}$ for $t \in [1, 4]$.

$\mathbf{r}'(t) = \frac{1}{t}\mathbf{i} - 2t\mathbf{j} + 2\mathbf{k}.$
$|\mathbf{r}'(t)| = \sqrt{\frac{1}{t^2} + 4t^2 + 4} = \sqrt{\frac{1+4t^2+4t^4}{t^2}}$

$= \sqrt{\frac{(1+2t^2)^2}{t^2}} = \frac{1+2t^2}{t} = \frac{1}{t} + 2t.$
The length of the arc is
$s = \int_1^4 \left(\frac{1}{t} + 2t\right) dt = \left.(\ln t + t^2)\right|_1^4$
$= 15 + \ln 4.$

2) Reparametrize
$\mathbf{r}(t) = (1 - t^2)\mathbf{i} + t^2\mathbf{j} + \sqrt{2}t^2\mathbf{k}$
in terms of the arc length function
from $a = 0$.

$\mathbf{r}'(t) = -2t\mathbf{i} + 2t\mathbf{j} + 2\sqrt{2}t\mathbf{k}.$
$|\mathbf{r}'(t)| = \sqrt{(-2t)^2 + (2t)^2 + (2\sqrt{2}t)^2} = 4t$
$s(t) = \int_0^t |\mathbf{r}'(u)|\,du = \int_0^t 4u\,du = 2t^2.$
From $s = 2t^2$, $t^2 = \frac{s}{2}$ and $t = \sqrt{\frac{s}{2}}$.
Thus the reparametrization is
$\mathbf{r}(t(s)) = \left(1 - \frac{s}{2}\right)\mathbf{i} + \frac{s}{2}\mathbf{j} + \frac{s}{\sqrt{2}}\mathbf{k}.$

B. For a smooth function $\mathbf{r}(t)$ the <u>unit</u>
<u>tangent vector</u> **T** is $\mathbf{T}(t) = \frac{\mathbf{r}'(t)}{|\mathbf{r}'(t)|}$ and
the principal <u>unit normal vector</u> **N** is
$\mathbf{N} = \frac{\mathbf{T}'}{|\mathbf{T}'|}.$
N is orthogonal to **T**.
The binormal vector **B** is
$\mathbf{B}(t) = \mathbf{T}(t) \times \mathbf{N}(t).$

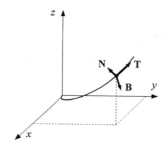

The plane formed by **N** and **B** is the <u>normal plane</u>; all vectors in the normal
plane are orthogonal to **T**. The plane formed by **T** and **N** is the <u>osculating</u>
<u>plane</u>; it is the plane that best approximates the direction of the curve.

The curvature κ for C for a twice differentiable $\mathbf{r}(t)$ is
$$\kappa = \left|\frac{d\mathbf{T}}{ds}\right| = \frac{|\mathbf{T}'(t)|}{|\mathbf{r}'(t)|} = \frac{|\mathbf{r}'(t) \times \mathbf{r}''(t)|}{|\mathbf{r}'(t)|^3}.$$
In the special case of $y = f(x)$ in the plane this becomes
$$\kappa = \frac{|y''|}{[1+(y')^2]^{3/2}}.$$

3) Find the principal unit normal vector
for $\mathbf{r}(t) = \sin t\mathbf{i} + 2t\mathbf{j} - \cos t\mathbf{k}$ at
$t = \frac{\pi}{4}$.

$\mathbf{r}'(t) = \cos t\mathbf{i} + 2\mathbf{j} + \sin t\mathbf{k}.$
$|\mathbf{r}'(t)| = \sqrt{\cos^2 t + 4 + \sin^2 t} = \sqrt{5}.$
Thus $\mathbf{T} = \frac{1}{\sqrt{5}}(\cos t\mathbf{i} + 2\mathbf{j} + \sin t\mathbf{k}).$
$\mathbf{T}' = \frac{1}{\sqrt{5}}(-\sin t\mathbf{i} + \cos t\mathbf{k}).$
$|\mathbf{T}'| = \frac{1}{\sqrt{5}}\sqrt{\sin^2 t + \cos^2 t} = \frac{1}{\sqrt{5}}.$
Thus $\mathbf{N} = \frac{\mathbf{T}'}{|\mathbf{T}'|} = -\sin t\mathbf{i} + \cos t\mathbf{k}.$
At $t = \frac{\pi}{4}$, $\mathbf{N} = -\frac{\mathbf{i}}{\sqrt{2}} + \frac{\mathbf{k}}{\sqrt{2}}.$

4) For $\mathbf{r}(t) = \frac{1}{2}t^3\mathbf{i} - 4t^2\mathbf{j} + 3t\mathbf{k}$ find

a) the unit tangent $\mathbf{T}$ at $t = 2$

$\mathbf{T}(t) = \mathbf{r}'(t) = \frac{3}{2}t^2\mathbf{i} - 8t\mathbf{j} + 3\mathbf{k}$.
$\mathbf{T}(2) = 6\mathbf{i} - 16\mathbf{j} + 3\mathbf{k}$.

b) the unit normal $\mathbf{N}$ at $t = 2$

$\mathbf{T}'(t) = 3t\mathbf{i} - 8\mathbf{j}$
$\mathbf{T}'(2) = 6\mathbf{i} - 8\mathbf{j}$
$|\mathbf{T}'(2)| = \sqrt{6^2 + (-8)^2 + 0^2} = 10$
$\mathbf{N}(2) = \frac{6}{10}\mathbf{i} - \frac{8}{10}\mathbf{j} = 0.6\mathbf{i} - 0.8\mathbf{j}$

c) the binormal $\mathbf{B}$ at $t = 2$

$$\mathbf{B}(2) = \mathbf{T}(2) \times \mathbf{N}(2) = \begin{vmatrix} \mathbf{i} & \mathbf{j} & \mathbf{k} \\ 6 & -16 & 3 \\ 0.6 & -0.8 & 0 \end{vmatrix}$$
$$= 2.4\mathbf{i} + 1.8\mathbf{j} + 4.8\mathbf{k}.$$

d) the normal plane at $t = 2$

The normal plane contains vectors
orthogonal to $\mathbf{T} = \langle 6, -16, 6 \rangle$.
At $t = 2$, $\mathbf{r}(2) = \langle 4, -16, 6 \rangle$.
$\mathbf{T} \cdot \mathbf{r}(2) = 24 + 256 + 18 = 298$.
The equation of the plane is
$6(x - 4) - 16(y + 16) + 3(z - 6) = 298$,
or $6x - 16y + 3z = 596$.

e) the osculating plane at $t = 2$

This plane contains vectors orthogonal to
$\mathbf{B} = \langle 2.4, 1.8, 4.8 \rangle$.
$\mathbf{B} \cdot \mathbf{r}(2) = 2.4(4) + 1.8(-16) + 4.8(6)$
$\qquad = 9.6$.
The equation of the plane is
$2.4(x - 4) + 1.8(y + 16) + 4.8(z - 6)$
$\qquad = 9.6$, or
$2.4x + 1.8y + 4.8z = 19.2$.

5) Find the curvature at $t = 1$ for the curve $\mathbf{r}(t) = t^2\mathbf{i} + \frac{2}{3}t^3\mathbf{j} + 2t\mathbf{k}$.

$\mathbf{r}'(t) = 2t\mathbf{i} + 2t^2\mathbf{j} + 2\mathbf{k}$.
$|\mathbf{r}'(t)| = \sqrt{4t^2 + 4t^4 + 4} = 2\sqrt{t^2 + t^4 + 1}$
$\mathbf{r}''(t) = 2\mathbf{i} + 4t\mathbf{j}$.
$$\mathbf{r}'(t) \times \mathbf{r}''(t) = \begin{vmatrix} \mathbf{i} & \mathbf{j} & \mathbf{k} \\ 2t & 2t^2 & 2 \\ 2 & 4t & 0 \end{vmatrix}$$
$$= -8t\mathbf{i} + 4\mathbf{j} + 4t^2\mathbf{k}$$
Thus $|\mathbf{r}' \times \mathbf{r}''| = \sqrt{64t^2 + 16 + 16t^4}$
$$= 4\sqrt{4t^2 + 1 + t^4}.$$

6) Find the curvature of $y = x^2$ at $x = \sqrt{6}$.

Therefore $\kappa = \dfrac{|\mathbf{r}' \times \mathbf{r}''|}{|\mathbf{r}'|^3} = \dfrac{4\sqrt{4t^2+1+t^4}}{\left(2\sqrt{t^2+t^4+1}\right)^3}$

$= \dfrac{\sqrt{4t^2+1+t^4}}{2(t^2+t^4+1)^{3/2}}.$

At $t = 1$, $\kappa = \dfrac{\sqrt{4+1+1}}{2(1+1+1)^{3/2}} = \dfrac{\sqrt{6}}{2(3)^{3/2}}$

$= \dfrac{\sqrt{6}}{2(3)\sqrt{3}} = \dfrac{\sqrt{2}}{6}.$

For $y = x^2$, $y' = 2x$, $y'' = 2$ and

$\kappa = \dfrac{|2|}{(1+(2x)^2)^{3/2}}.$

At $x = \sqrt{6}$, $\kappa = \dfrac{2}{(25)^{3/2}} = \dfrac{2}{125}.$

Motion in Space: Velocity and Acceleration

Concepts to Master

A. Velocity; Speed; Acceleration; Newton's Second Law of Motion
B. Tangent and normal components of acceleration

Summary and Focus Questions

A. A vector function $\mathbf{r}(t)$ may be thought of as the position of a particle in space at time t.

   $\mathbf{v}(t) = \mathbf{r}'(t)$ is the <u>velocity</u> vector and $|\mathbf{r}'(t)|$ is the <u>speed</u>.
   $\mathbf{a}(t) = \mathbf{r}''(t)$ is the <u>acceleration</u> vector.

The unit tangent normal $\mathbf{T}$ points in the direction the particle is moving and the unit normal vector $\mathbf{N}$ points in a direction orthogonal to the motion. Vector integrals may be used to determine velocity $\mathbf{v}(t) = \int \mathbf{a}(t)\, dt$ and position $\mathbf{r}(t) = \int \mathbf{v}(t)\, dt$.

Newton's Second Law of Motion is $\mathbf{F}(t) = m\mathbf{a}(t)$, where $\mathbf{F}(t)$ is the force acting on an object of mass m whose acceleration is $\mathbf{a}(t)$.

In the particular case of a projectile of mass m fired at an angle α with initial velocity
$\mathbf{v}_0 = (\mathbf{v}_0 \cos \alpha)\mathbf{i} + (\mathbf{v}_0 \sin \alpha)\mathbf{j}$:

$\quad \mathbf{F} = -mg\mathbf{j} \quad (g = 9.8 \text{ m/sec}^2)$
$\quad \mathbf{a} = -g\mathbf{j}$
$\quad \mathbf{v} = -gt\mathbf{j} + \mathbf{v}_0$
$\quad \mathbf{r} = -\frac{1}{2}gt^2\mathbf{j} + t\mathbf{v}_0$

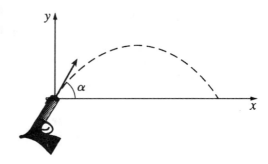

1) Find the velocity, speed, and acceleration of a particle whose position at time t is
$\mathbf{r}(t) = \langle 3t^2, \; t^3 + 1, \; e^{-t} \rangle$.

$\mathbf{v}(t) = \mathbf{r}'(t) = \langle 6t, \; 3t^2, \; -e^{-t} \rangle.$
Speed is $|\mathbf{v}(t)| = \sqrt{36t^2 + 9t^4 + e^{-2t}}.$
$\mathbf{a}(t) = \langle 6, \; 6t, \; e^{-t} \rangle.$

2) Find the velocity and position at time t of a particle whose initial position is $\mathbf{i} + \mathbf{j}$, initial velocity is $\mathbf{j} + \mathbf{k}$, and acceleration is $\mathbf{a}(t) = 12t\mathbf{i} + 2\mathbf{k}$.

$\mathbf{v}(t) = \int \mathbf{a}(t)dt = 6t^2\mathbf{i} + 2t\mathbf{k} + \mathbf{C}.$
Initially $(t = 0)$ $\mathbf{v}_0 = \mathbf{j} + \mathbf{k}.$
$0\mathbf{i} + 0\mathbf{k} + \mathbf{C} = \mathbf{j} + \mathbf{k}.$
$\mathbf{v}(t) = 6t^2\mathbf{i} + 2t\mathbf{k} + (\mathbf{j} + \mathbf{k})$
$\qquad = 6t^2\mathbf{i} + \mathbf{j} + (2t + 1)\mathbf{k}.$
$\mathbf{r}(t) = \int \mathbf{v}(t)dt = 2t^3\mathbf{i} + t\mathbf{j} + (t^2 + t)\mathbf{k} + \mathbf{D}$
Initially, $\mathbf{r}(0) = \mathbf{i} + \mathbf{j}:$
$0\mathbf{i} + 0\mathbf{j} + 0\mathbf{k} + \mathbf{D} = \mathbf{i} + \mathbf{j}$
$\mathbf{r}(t) = 2t^3\mathbf{i} + t\mathbf{j} + (t^2 + t)\mathbf{k} + \mathbf{i} + \mathbf{j}$
$\qquad = (2t^3 + 1)\mathbf{i} + (t + 1)\mathbf{j} + (t^2 + t)\mathbf{k}.$

3) What force is required in order for a 5 kg mass to be pushed in such a way that, at time t, its position is $\mathbf{r}(t) = 3t^2\mathbf{i} + t^4\mathbf{j} + 2t^3\mathbf{k}$?

$\mathbf{v}(t) = \mathbf{r}'(t) = 6t\mathbf{i} + 4t^3\mathbf{j} + 6t^2\mathbf{k}.$
$\mathbf{a}(t) = \mathbf{v}'(t) = 6\mathbf{i} + 12t^2\mathbf{j} + 12t\mathbf{k}.$
$\mathbf{F}(t) = 5\mathbf{a}(t) = 30\mathbf{i} + 60t^2\mathbf{j} + 60t\mathbf{k}.$

4) A projectile is fired with a velocity of 200 m/sec at an inclination of 30° from a point 10 meters above ground. Find the vector function that describes the path of motion.

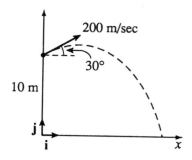

Choose a coordinate system so that from Newton's Second Law, $\mathbf{a}(t) = -g\mathbf{j}$.
Thus $\mathbf{v}(t) = \int -g\mathbf{j}\ dt = -gt\mathbf{j} + \mathbf{C}$.
At $t = 0$, $\mathbf{v}_0 = \mathbf{C}$. Since the velocity is 200 m/sec at a 30° inclination,
$\mathbf{C} = (200\cos 30°)\mathbf{i} + (200\sin 30°)\mathbf{j}$
$\qquad = 100\sqrt{3}\mathbf{i} + 100\mathbf{j}$.
Thus $\mathbf{v}(t) = 100\sqrt{3}\mathbf{i} + (100 - gt)\mathbf{j}$.
$\mathbf{r}(t) = \int \mathbf{v}(t)dt$
$\qquad = 100\sqrt{3}t\mathbf{i} + \left(100t - \frac{gt^2}{2}\right)\mathbf{j} + \mathbf{D}$.
At $t = 0$, $\mathbf{r}(0) = \mathbf{D}$. Since the projectile is 10 meters above ground, $\mathbf{D} = 10\mathbf{j}$. Thus
$\mathbf{r}(t) = 100\sqrt{3}t\mathbf{i} + \left(100t - \frac{gt^2}{2} + 10\right)\mathbf{j}$.

B. The acceleration vector **a** lies in the plane determined by the unit tangent vector **T** and the unit normal vector **N**. Thus **a** may be written as a combination of **T** and **N**:

$\qquad \mathbf{a} = a_T\mathbf{T} + a_N\mathbf{N}$,

where the tangential component is $a_T = v'$ (v is the speed, $v = |\mathbf{v}|$).
and the normal component is $a_N = \kappa v^2$ (κ is the curvature).

In terms of the position function $\mathbf{r}(t)$,

$\qquad a_T = \frac{\mathbf{r}' \cdot \mathbf{r}''}{|\mathbf{r}'|}$ and $a_N = \frac{|\mathbf{r}' \times \mathbf{r}''|}{|\mathbf{r}'|}$.

5) Find the tangential and normal components of acceleration for $\mathbf{r}(t) = t^2\mathbf{i} + t^4\mathbf{j} + t^3\mathbf{k}$.

$\mathbf{r}'(t) = 2t\mathbf{i} + 4t^3\mathbf{j} + 3t^2\mathbf{k}$.
$|\mathbf{r}'(t)| = \sqrt{4t^2 + 16t^6 + 9t^4}$
$\qquad = t\sqrt{4 + 16t^4 + 9t^2}$.
$\mathbf{r}''(t) = 2\mathbf{i} + 12t^2\mathbf{j} + 6t\mathbf{k}$.
$\mathbf{r}' \cdot \mathbf{r}'' = 4t + 48t^5 + 18t^3$.

$\mathbf{r}' \times \mathbf{r}'' = \begin{vmatrix} \mathbf{i} & \mathbf{j} & \mathbf{k} \\ 2t & 4t^3 & 3t^2 \\ 2 & 12t^2 & 6t \end{vmatrix}$
$\qquad = -12t^4\mathbf{i} - 6t^2\mathbf{j} + 16t^3\mathbf{k}$.

$|\mathbf{r}' \times \mathbf{r}''| = \sqrt{144t^8 + 36t^4 + 256t^6}$

$\qquad = 2t^2\sqrt{36t^4 + 9 + 64t^2}$.

Thus $a_T = \frac{4t + 48t^5 + 18t^3}{t\sqrt{4 + 16t^4 + 9t^2}} = \frac{4 + 48t^4 + 18t^2}{\sqrt{4 + 16t^4 + 9t^2}}$

and

$a_N = \frac{2t^2\sqrt{36t^4 + 9 + 64t^2}}{t\sqrt{4 + 16t^4 + 9t^2}} = \frac{2t\sqrt{36t^4 + 9 + 64t^2}}{\sqrt{4 + 16t^4 + 9t^2}}$.

Cylindrical and Spherical Coordinates

Concepts to Master

Cylindrical coordinates; Spherical coordinates; Conversion from one set of coordinates to another.

Summary and Focus Questions

The <u>cylindrical coordinates</u> (r, θ, z) of a point P in space are polar coordinates (r, θ) in the xy-plane and the usual z coordinate. They are called cylindrical because $r = a$ is a circular cylinder about the z-axis.

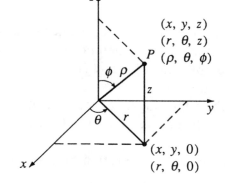

The <u>spherical coordinates</u> (ρ, θ, ϕ) of a point P in space are defined as:

ρ = distance from P to the origin.
θ = same as cylindrical coordinates.
ϕ = angle between the positive z-axis and the line from P to the origin.

They are called spherical because $\rho = a$ is a sphere about the origin.

Thus we have three different coordinate systems for three-dimensional space. The following table shows how to convert coordinates for one system to the others.

Conversion from <u>Rectangular</u> (x, y, z) to

<u>Cylindrical</u> (r, θ, z)	<u>Spherical</u> (ρ, θ, ϕ)
$r^2 = x^2 + y^2$	$\rho^2 = x^2 + y^2 + z^2$
$\tan \theta = \frac{y}{x}$	$\tan \theta = \frac{y}{x}$
$z = z$	$\cos \phi = \frac{z}{\rho}$

Conversion from <u>Cylindrical</u> (r, θ, z) to

<u>Rectangular</u> (x, y, z) <u>Spherical</u> (ρ, θ, ϕ)

$x = r \cos \theta$ $\rho^2 = r^2 + z^2$

$y = r \sin \theta$ $\theta = \theta$

$z = z$ $\cos \phi = \frac{z}{\rho}$

Conversion from <u>Spherical</u> (ρ, θ, ϕ) to

<u>Rectangular</u> (x, y, z) <u>Cylindrical</u> (r, θ, z)

$x = \rho \sin \phi \cos \theta$ $r = \rho \sin \phi$

$y = \rho \sin \phi \sin \theta$ $\theta = \theta$

$z = \rho \cos \phi$ $z = \rho \cos \phi$

1) Find the indicated coordinates of each point:

a) Spherical: $\left(8, \frac{\pi}{3}, \frac{\pi}{6}\right)$.
   Rectangular: _____.

$(2, 2\sqrt{3}, 4\sqrt{3})$.
$x = 8 \sin \frac{\pi}{6} \cos \frac{\pi}{3} = 8 \cdot \frac{1}{2} \cdot \frac{1}{2} = 2.$
$y = 8 \sin \frac{\pi}{6} \sin \frac{\pi}{3} = 2\sqrt{3}.$
$z = 8 \cos \frac{\pi}{6} = 8 \frac{\sqrt{3}}{2} = 4\sqrt{3}.$

b) Cylindrical: $\left(\sqrt{3}, \frac{\pi}{6}, 1\right)$.
   Spherical: _____.

$\left(2, \frac{\pi}{6}, \frac{\pi}{3}\right)$.
$\rho^2 = \sqrt{3}^2 + 1^2 = 4$, so $\rho = 2.$
$\theta = \frac{\pi}{6}.$
$\cos \phi = \frac{1}{2}$, so $\phi = \frac{\pi}{3}.$

c) Rectangular: $(2, 2\sqrt{3}, 6)$.
   Cylindrical: _____.

$\left(4, \frac{\pi}{3}, 6\right)$.
$r^2 = 2^2 + (2\sqrt{3})^2 = 16$, $r = 4.$
$\tan \theta = \frac{2\sqrt{3}}{2} = \sqrt{3}, \theta = \frac{\pi}{3}.$
$z = 6.$

d) Spherical: $\left(10, \frac{\pi}{6}, \frac{\pi}{3}\right)$.
   Cylindrical: _____.

$\left(5\sqrt{3}, \frac{\pi}{6}, 5\right)$.
$r = 10 \sin \frac{\pi}{3} = 10\left(\frac{\sqrt{3}}{2}\right) = 5\sqrt{3}.$
$\theta = \frac{\pi}{6}.$
$z = 10 \cos \frac{\pi}{3} = 10\left(\frac{1}{2}\right) = 5.$

2) Identify the surface with equation:

    a) $\rho = \cos\phi$ in spherical coordinates.

Switch to rectangular coordinates.
$$\rho = \cos\phi$$
$$\rho^2 = \rho\cos\phi$$
$$x^2 + y^2 + z^2 = z$$
$$x^2 + y^2 + \left(z - \tfrac{1}{2}\right) = \tfrac{1}{4}.$$
This is a sphere with radius $\tfrac{1}{2}$ and center $\left(0, 0, \tfrac{1}{2}\right)$.

    b) $z^2 = r^2$ in cylindrical coordinates.

Switch to rectangular: $z^2 = x^2 + y^2$.
This is an elliptic cone.

12

Partial Derivatives

"WHAT'S MOST DEPRESSING IS THE REALIZATION THAT EVERYTHING WE BELIEVE WILL BE DISPROVED IN A FEW YEARS."

Cartoons courtesy of Sidney Harris. Used by permission.

Section 12.1

Functions of Several Variables

Concepts to Master

Functions of 2, 3, ..., n variables; Domain; Range; Graphs; Level curves

Summary and Focus Questions

A <u>function of two variables</u> assigns to each ordered pair (x, y) in $D \subset R^2$ a unique real number $f(x, y)$. D is the <u>domain</u> of f. The <u>range</u> of f is the set of all $f(x, y)$ values.

There are three different ways to view the domain of f.

1. f is a function of two independent variables, x and y, where $(x, y) \in D$.
2. f is a function whose domain is all points (x, y) in D.
3. f is a function whose domain is all vectors $\langle x, y \rangle$ in D.

For example, let $f(x, y) = x^2 + y^2$. Then for $x = 2$, $y = 1$
$f(2, 1) = 2^2 + 1^2 = 5$. The value 5 may be thought of as being associated with the values $x = 2$ and $y = 1$, or with the point $(2, 1)$, or with the vector $\langle 2, 1 \rangle$.

The graph of a function of two variables is a surface in three dimensional space. (x, y, z) is on the graph of f if and only if $z = f(x, y)$.

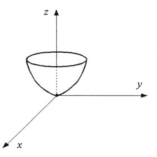

Our example, $f(x, y) = x^2 + y^2$ has a graph which is a circular paraboloid.

For a constant k, the <u>level curve determined by k</u> is $\{(x, y) \in D \mid f(x, y) = k\}$.

Each level curve is a subset of the domain of f. Level curves are not part of the graph of f. Level curves for various values of k may be drawn to obtain a visualization of the graph of $z = f(x, y)$, much in the same fashion as you see temperature regions on a weather map.

For our example $f(x, y) = x^2 + y^2$,
each $k > 0$ produces the level curve
$\{(x, y) | x^2 + y^2 = k\}$ which is a circle
in the xy-plane.

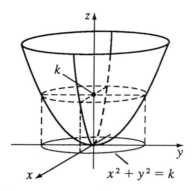

$x^2 + y^2 = k$

A function of three variables $f(x, y, z)$
has a graph in 4-dimensional space and
level surfaces that are subsets of R^3.
Functions of more than three variables
are defined similarly.

1. What is the domain of
    $f(x, y) = \frac{1}{5x-10y}$?

$f(x, y)$ is defined for all (x, y) except
when $5x - 10y = 0$ or $x = 2y$. The domain
is $\{(x, y) | x \neq 2y\}$.

2. What are the domain and range of
    $f(x, y) = \frac{1}{x^2+y^2}$?

$x^2 + y^2 = 0$ only for $(x, y) = (0, 0)$.
The domain is $\{(x, y) | x \neq 0 \text{ and } y \neq 0\}$.
$x^2 + y^2 > 0$ for all $(x, y) \neq (0, 0)$.
Thus the range is $(0, \infty)$.

3. Sketch the level curves for
    $f(x, y) = x - y^2$ for $k = 0, 2, -2, 4$.

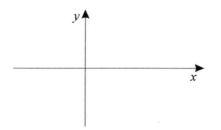

Each level curve has the form $x - y^2 = k$
which is a parabola opening to the right
with vertex $(k, 0)$.

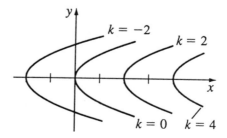

4) Describe the graph of $f(x, y) = x^2$.

$z = x^2$ is a parabolic cylinder along the y-axis:

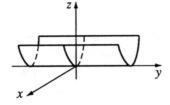

5) True or False:
    Two different level curves can never intersect.

True, since every point in the domain has a unique functional value.

6) Describe the level surfaces of
    $f(x, y, z) = x^2 + 4y^2 + 9z^2$.

For any $k \geq 0$, the level surface is all (x, y, z) such that $x^2 + 4y^2 + 9z^2 = k$, which is an ellipsoid with center $(0, 0, 0)$.

Limits and Continuity

Concepts to Master

A. Limits of functions of two or more variables
B. Continuity of functions of two or more variables

Summary and Focus Questions

A. Let f be defined on a disk with center (a, b) except perhaps at (a, b). Then
$$\lim_{(x,\, y)\to(a,\, b)} f(x,\, y) = L$$
means for all $\varepsilon > 0$ there exists $\delta < 0$ such that $\left|f(x,\, y) - L\right| < \varepsilon$ whenever $0 < \sqrt{(x-a)^2 + (y-b)^2} < \delta$.

Using $\mathbf{x} = \langle x,\, y \rangle$ and $\mathbf{a} = \langle a,\, b \rangle$ this may be rewritten as for all $\varepsilon > 0$ there exists $\delta > 0$ such that
$$\left|f(\mathbf{x}) - L\right| < \varepsilon \text{ whenever } 0 < \left|\mathbf{x} - \mathbf{a}\right| < \delta.$$
This version definition extends to functions of three or more variables without modification. Similar limit theorems like those for functions of one variable hold; for example,

if $\displaystyle\lim_{(x,y)\to(a,b)} f(x,\, y) = L$ and $\displaystyle\lim_{(x,y)\to(a,b)} g(x,\, y) = M$, then
$$\lim_{(x,y)\to(a,b)} [f(x,\, y) + g(x,\, y)] = L + M.$$

If $\displaystyle\lim_{(x,y)\to(a,b)} f(x,\, y) = L$, then $f(x,\, y)$ approaches L as $(x,\, y)$ approaches $(a,\, b)$ along any curve containing $(a,\, b)$. If two different curves containing $(a,\, b)$ yield different limits as $(x,\, y)$ approaches $(a,\, b)$ along each, then the limit of f does not exist.

1) Evaluate each:

   a) $\displaystyle\lim_{(x,y)\to(4,\, 2)} (6x + 3y)$

   $6(4) + 3(2) = 30.$

b) $\lim\limits_{(x,\,y)\to(0,\,0)} e^{1/(x^2+y^2)}$

$\infty.$

c) $\lim\limits_{(x,\,y)\to(3,\,2)} (y^3 - y)$

$2^3 - 2 = 6.$

d) $\lim\limits_{(x,\,y)\to(1,\,-1)} \frac{x^2-2xy+y^2-4}{x-y-2}$

At $(1,\,-1)$ we have $\frac{0}{0}$. We can factor and cancel: $\frac{x^2-2xy+y^2-4}{x-y-2} = \frac{(x-y)^2-4}{x-y-2}$

$$= \frac{(x-y+2)(x-y-2)}{x-y-2} = x - y + 2$$
$$\lim\limits_{(x,\,y)\to(1,\,-1)} (x - y + 2) = 1 + 1 + 2 = 4.$$

2)   Let $f(x,\,y) = \frac{x^3y}{2x^6+y^2}$
for $(x,\,y) \neq (0,\,0)$.
Evaluate $\lim\limits_{(x,\,y)\to(0,\,0)} f(x,\,y):$

a)   along the curve $y = 0$.

$f(x,\,0) = 0$ for all x so f has limit 0 along $y = 0$.

b)   along the curve $y = x^2$.

$f(x,\,x^2) = \frac{x^3 x^2}{2x^6+x^4} = \frac{x}{2x^2+1}.$

As $x \to 0$, $\frac{x}{2x^2+1} \to 0$, so f has limit 0 along $y = x^2$.

c)   along the curve $y = x^3$.

$f(x,\,x^3) = \frac{x^3 x^3}{2x^6+x^6} = \frac{1}{3}.$

f has limit $\frac{1}{3}$ along $y = x^3$.

3)   Find $\lim\limits_{(x,\,y)\to(0,\,0)} f(x,\,y)$ for the function in question 2.

By parts b) and c) f approaches different limits along different curves. Thus the limit does not exist.

4)   True, False:
If $\lim\limits_{(x,\,y)\to(a,\,b)} f(x,\,y) = L$ and
$\lim\limits_{(x,\,y)\to(a,\,b)} g(x,\,y) = M$ then
$\lim\limits_{(x,\,y)\to(a,\,b)} f(x,\,y)g(x,\,y) = LM.$

True.

B. The function $z = f(x, y)$ <u>is continuous at (a, b)</u> if
$$\lim_{(x,y)\to(a,b)} f(x, y) = f(a, b).$$

Polynomials in x and y are continuous everywhere; rational functions are continuous everywhere they are defined.

5) Where is $f(x, y) = \ln(x - y)$ continuous?

$\ln(x - y)$ is continuous on its domain. The domain is $\{(x, y) \mid x > y\}$.

6) Where is $f(x, y) = \frac{2x+y}{x^2+xy}$ continuous?

Since f is a rational function, f is continuous everywhere on its domain. The domain is all points (x, y) such that $x^2 + xy \neq 0$, i.e., $x(x + y) \neq 0$ so $x \neq 0$ and $x \neq -y$.

7) Is
$$f(x, y) = \begin{cases} \frac{x^3 y}{2x^6+y^2} & (x, y) \neq (0, 0) \\ 0 & (x, y) = (0, 0) \end{cases}$$
continuous at $(0, 0)$?

No. By problem 3) $\lim\limits_{(x,y)\to(0,0)} f(x, y)$ does not exist.

Partial Derivatives

Concepts to Master

A. Partial derivatives; Slopes of tangent line; Instantaneous rate of change; Implicit partial differentiation

B. Higher order partial derivatives; Clairaut's Theorem

Summary and Focus Questions

A. For $z = f(x, y)$, the partial derivative of f with respect to x is

$$f_x(x, y) = \lim_{h \to 0} \frac{f(x+h, y) - f(x, y)}{h}.$$

Since y is unchanging in this definition, $f_x(x, y)$ is computed by treating y as a constant. For example, if $f(x, y) = x^3 y^2$, then
$f_x(x, y) = (3x^2)y^2 = 3x^2 y^2$.
Other notations for $f_x(x, y)$ include

$$f_x, \frac{\partial f}{\partial x}, \frac{\partial}{\partial x} f(x, y), \frac{\partial z}{\partial x}, D_x f, D_x f(x, y).$$

$f_x(a, b)$ may be interpreted as the slope of the tangent line to the surface $z = f(x, y)$ determined by the trace $y = b$.

$f_x(x, y)$ is the instantaneous rate of change of f with respect to x.

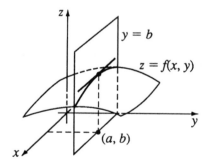

Likewise $f_y(x, y) = \lim_{h \to 0} \frac{f(x, y+h) - f(x, y)}{h}$ is computed by holding x constant and is the slope of the tangent line to $z = f(x, y)$ in the y direction.
In a similar manner, partial derivatives may be defined and computed for functions of more than two variables. For example, if

$$f(x, y, z, w) = 2x^2 y + 3xz^3 w + z^2 y^2 w,$$

92

then, treating x, y, and w as constants, we compute
$$\frac{\partial f}{\partial z} = 0 + 3x(3z^2)w + (2z)(y^2w) = 9xz^2w + 2zy^2w.$$

Implicit partial differentiation is performed in the same manner as was done for functions of one variable but remember to treat the other variables as constants. For example, to compute $\frac{\partial f}{\partial x}$ where $z = f(x, y)$ is defined by $4x^2 + 9y^2 + 16z^2 = 100$, $8x + 0 + 32z\frac{\partial z}{\partial x} = 0$. Hence $\frac{\partial z}{\partial x} = -\frac{x}{4z}$.

1) For $w = f(x, y, z)$ define $\frac{\partial f}{\partial z}$.

$$\frac{\partial f}{\partial z} = \lim_{h \to 0} \frac{f(x, y, z+h) - f(x, y, z)}{h}.$$

2) Find f_x and f_y for

a) $f(x, y) = e^{x^2 - y^2}$

$f_x = e^{x^2 - y^2}(2x) = 2xe^{x^2 - y^2}$.
$f_y = e^{x^2 - y^2}(-2y) = -2ye^{x^2 - y^2}$.

b) $f(x, y) = xy \sec x$

$f_x = (xy)(\sec x \tan x) + (\sec x)y$
$\quad = y \sec x(x \tan x + 1)$.
$f_y = x \sec x$.

3) Find f_y for $f(x, y, z) = \frac{x}{\sqrt{y+z}}$.

$f(x, y, z) = x(y + z)^{-1/2}$.
$f_y = x\left(-\frac{1}{2}\right)(y + z)^{-3/2}(1) = -\frac{x}{2(y+z)^{3/2}}$.

4) Find the slope of the tangent line in the y direction to $f(x, y) = x^2 + 4xy + y^2$ at the point $(1, 2, 13)$.

$f_y = 4x + 2y$.
$f_y(1, 2) = 4(1) + 2(2) = 8$.

5) Find $\frac{\partial z}{\partial x}$ where $z = f(x, y)$ is given by $x^2 + y^2 + z^2 = \sin(xyz)$.

Treat y as a constant:
$$2x + 0 + 2z \cdot \frac{\partial z}{\partial x} = \cos(xyz)\left[xy\frac{\partial z}{\partial x} + yz\right].$$
$$2x - yz\cos(xyz) = \frac{\partial z}{\partial x}[xy\cos(xyz) - 2z]$$
$$\frac{\partial z}{\partial x} = \frac{2x - yz\cos(xyz)}{xy\cos(xyz) - 2z}.$$

B. For a function $z = f(x, y)$ there are four second partial derivatives:

$$f_{xx}(x, y) = \frac{\partial^2 f}{\partial x^2} \qquad\qquad f_{yy}(x, y) = \frac{\partial^2 f}{\partial y^2}$$

$$f_{xy}(x, y) = \frac{\partial^2 f}{\partial y \partial x} \qquad\qquad f_{yx}(x, y) = \frac{\partial^2 f}{\partial x \partial y}$$

For example, $f_{xy}(x, y)$ is determined by finding the partial derivative of f with respect to x and taking the derivative of that result with respect to y.

<u>Clairaut's Theorem</u>: If f is defined on a disk containing (a, b) and both f_{xy} and f_{yx} are continuous on that disk, then $f_{xy} = f_{yx}$ at (a, b).

6) Find $\frac{\partial^2 f}{\partial x \partial y}$ for $f(x, y) = 3xy^3 + 8x^2y^4$

$f_y = 9xy^2 + 32x^2y^3$.

Thus $f_{yx} = \frac{\partial^2 f}{\partial x \partial y} = 9y^2 + 64xy^3$.

7) For $f(x, y) = \sin(x^2 + y^2)$, is $f_{xy} = f_{yx}$ for all (x, y)?

Yes, because both f_{xy} and f_{yx} are continuous.

8) True or False:
    If all third partial derivatives are continuous, then

    a) $f_{xyz} = f_{zxy}$

    True.

    b) $f_{xxyz} = f_{xyzy}$

    False.

9) Let $f(x, y) = 4x^2y^5 + 3x^3y^2$

    a) Compute f_{xyy}.

    $f_x = 8xy^5 + 9x^2y^2$
    $f_{xy} = 40xy^4 + 18x^2y$
    $f_{xyy} = 160xy^3 + 18x^2$.

    b) Compute f_{yxy}.

    $f_{yxy} = 160xy^3 + 18x^2$ (same as part (a)).

Tangent Planes and Differentials

Concepts to Master

A. Tangent plane to $z = f(x, y)$
B. Differential; Differentiability

Summary and Focus Questions

A. For a surface given by $z = f(x, y)$, where f has
continuous first partial derivatives, all the tangent
lines at a given point to a surface form a plane
called the <u>tangent plane</u>. If $P: (x_0, y_0, z_0)$ is a
point on $z = f(x, y)$, the tangent plane at P has
equation

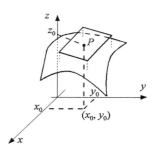

$$z - z_0 = f_x(x_0, y_0)(x - x_0) + f_y(x_0, y_0)(y - y_0).$$

This equation has some familiar terms in it.
For example, $z - z_0 = f_x(x_0, y_0)$ is the equation of the tangent line to the curve
$z = f(x, y_0)$ (y_0 is constant) in the plane $y = y_0$.

1) Find the equation of the tangent plane
   to $z = 4x^2y$ at $(1, 3)$.

$f(1, 3) = 4(1)^2 3 = 12.$
$f_x = 8xy, \; f_x(1, 3) = 8(1)(3) = 24.$
$f_y = 4x^2, \; f_y(1, 3) = 4(1)^2 = 4.$
The plane is
$z - 12 = 24(x - 1) + 4(y - 3)$, or
$24x + 4y - z = 24.$

B. If $z = f(x, y)$ then, similar to functions of one variable, the <u>change in z</u> is
defined as $\Delta z = f(x + \Delta x, y + \Delta y) - f(x, y).$

If we let $dx = \Delta x$ and $dy = \Delta y$ then the <u>(total) differential</u> is
$$dz = f_x(x, y)dx + f_y(x, y)dy.$$

95

The function $z = f(x, y)$ is <u>differentiable at (a, b)</u> if
$$\Delta z = f_x(a, b)\Delta x + f_y(a, b)\Delta y + \varepsilon_1 \Delta x + \varepsilon_2 \Delta y$$
where $\varepsilon_1, \varepsilon_2$ are each functions of Δx and Δy such that $\varepsilon_1 \to 0$ and $\varepsilon_2 \to 0$ as $(\Delta x, \Delta y) \to 0$.

The functions ε_1 and ε_2 measure the difference between Δz and dz. Since their limits are 0 when f is differentiable, dz may be used to approximate Δz.

Similar results hold for functions of three or more variables.

2) Find Δz and dz for $z = 2x^2 + y^3$ as (x, y) changes from $(2, 1)$ to $(2.01, 1.03)$.

At $(2, 1)$, $z = 2(2)^2 + 1^3 = 9$.
At $(2.01, 1.03)$,
$z = 2(2.01)^2 + (1.01)^3 = 9.1729$.
Thus $\Delta z = 9.1729 - 9 = 0.1729$.
$\frac{\partial z}{\partial x} = 4x$. At $(2, 1)$, $\frac{\partial z}{\partial x} = 4(2) = 8$.

$\frac{\partial z}{\partial y} = 3y^2$. At $(2, 1)$, $\frac{\partial z}{\partial y} = 3(1)^2 = 3$.
$dx = \Delta x = 2.01 - 2 = 0.01$
$dy = \Delta y = 1.03 - 1 = 0.03$
Thus $dz = f_x dx + f_y dy$
$$= 8(0.01) + 3(0.03) = 0.17.$$
So dz is within 0.0029 of Δz.

3) Find dz for $z = xe^{xy}$.

$dz = (xye^{xy} + e^{xy})dx + (x^2 e^{xy})dy.$

4) For $w = f(x, y, z) = xy^3 + yz^3$, find dw.

$dw = f_x dx + f_y dy + f_z dz$
$$= y^3 dx + (3xy^2 + z^3)dy + 3yz^2 dz.$$

5) Estimate $\frac{(3.02)^2}{(0.99)^3}$ using a differential.

Let $z = f(x, y) = x^2 y^{-3}$. We estimate $f(3.02, 0.99)$ by calculating $f(3, 1)$ and dz.
$f(3, 1) = 3^2(1)^{-3} = 9$.
$f_x = 2xy^{-3} = 6$ at $(3, 1)$.
$f_y = -3x^2 y^{-4} = -27$ at $(3, 1)$. Thus
$dz = f_x dx + f_y dy$
$$= 6(0.02) + (-27)(-0.01) = 0.39.$$
Finally, $f(3.02, 0.99) = f(3, 1) + \Delta z$
$$\approx f(3, 1) + dz = 9 + 0.39 = 9.39.$$
$\left(\text{Note: To 4 decimals, } \frac{(3.02)^2}{(0.99)^3} = 9.3995.\right)$

6) True or False:

    a) If $f_x(x_0, y_0)$ and $f_y(x_0, y_0)$ exist
        then f is differentiable at
        (x_0, y_0).

False. (f_x and f_y must also be continuous.)
A counterexample at $(x_0, y_0) = (0, 0)$ is
$$f(x, y) = \begin{cases} \frac{xy}{x^2+y^2} & (x, y) \neq (0, 0) \\ 0 & (x, y) = (0, 0) \end{cases}$$
Both f_x and f_y are zero at $(0, 0)$ but f is
not continuous.

    b) If f is differentiable at (x_0, y_0)
        then $f_x(x_0, y_0)$ and $f_y(x_0, y_0)$
        exist.

True.

7) A circular spa is 7 ft. in diameter and 3
    ft. deep. If the measurements are
    accurate to within 0.05 ft. then use
    differentials to estimate the maximum
    error in calculating the volume of w.

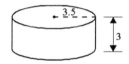

$V = \pi r^2 h$
$dV = \frac{\partial V}{\partial r} \cdot dr + \frac{\partial V}{\partial h} \cdot dh$
$\quad = 2\pi r h \; dr + \pi r^2 \; dh$
$\quad \approx 2\pi r h \Delta r + \pi r^2 \Delta h.$
We are given $|\Delta r| \leq 0.05$ and
$|\Delta h| \leq 0.05$. Thus, at $r = 3.5$ and $h = 3$
we have
$dV \approx 2\pi(3.5)(3)(0.05) + \pi(3.5)^2(0.05)$
$\quad = 1.6625\pi \approx 5.22$ ft.^{3}.

The Chain Rule

Concepts to Master

A. Forms of the Chain Rule
B. Implicit Differentiation using the Chain Rule

Summary and Focus Questions

A. <u>The Chain Rule</u>: Suppose x and y are differentiable functions of t and $z = f(x, y)$ is differentiable. Then z is differentiable with respect to t and
$$\frac{dz}{dt} = \frac{\partial z}{\partial x}\frac{dx}{dt} + \frac{\partial z}{\partial y}\frac{dy}{dt}.$$

Suppose $x = g(s, t)$, $y = h(s, t)$, and $z = f(x, y)$. The version of the Chain Rule in this case for computing the partial derivatives of z is
$$\frac{\partial z}{\partial s} = \frac{\partial z}{\partial x}\frac{\partial x}{\partial s} + \frac{\partial z}{\partial y}\frac{\partial y}{\partial s} \qquad\qquad \frac{\partial z}{\partial t} = \frac{\partial z}{\partial x}\frac{\partial x}{\partial t} + \frac{\partial z}{\partial y}\frac{\partial y}{\partial t}.$$

Each rule may be generalized. For example, for $w = f(x, y, z)$, $x = h(u, v, s, t)$, $y = k(u, v, s, t)$ and $z = m(u, v, s, t)$, the function w has four partial derivatives which may be found by the Chain Rule. For instance,
$$\frac{\partial w}{\partial v} = \frac{\partial w}{\partial x}\frac{\partial x}{\partial v} + \frac{\partial w}{\partial y}\frac{\partial y}{\partial v} + \frac{\partial w}{\partial z}\frac{\partial z}{\partial v}.$$
There are as many terms in the sum as there are intermediate variables (x, y, and z).

1) Find $\frac{dz}{dt}$ where $z = x^2y$, $x = e^t$,
   $y = t^2$.

   $\frac{dz}{dt} = \frac{\partial z}{\partial x}\frac{dx}{dt} + \frac{\partial z}{\partial y}\frac{dy}{dt} = (2xy)e^t + x^2(2t)$
   $= (2e^t t^2)e^t + (e^t)^2 2t = 2t^2 e^{2t} + 2te^{2t}.$

2) Find $\frac{\partial w}{\partial u}$ at $(u, v) = \left(\frac{\pi}{2}, \frac{\pi}{2}\right)$, where
   $w = x^2yz$, $x = uv$, $y = u\sin v$,
   $z = v\sin u$.

   $\frac{\partial w}{\partial u} = \frac{\partial w}{\partial x}\frac{\partial x}{\partial u} + \frac{\partial w}{\partial y}\frac{\partial y}{\partial u} + \frac{\partial w}{\partial z}\frac{\partial z}{\partial u}$
   $= (2xyz)v + x^2z(\sin v) + x^2y(v\cos u).$
   At $(u, v) = \left(\frac{\pi}{2}, \frac{\pi}{2}\right)$, $x = \frac{\pi^2}{4}$,
   $y = \frac{\pi}{2}$, $z = \frac{\pi}{2}.$

Therefore

$$\frac{\partial w}{\partial u} = \left(2\frac{\pi^2}{4}\frac{\pi}{2}\frac{\pi}{2}\right)\frac{\pi}{2} + \left(\frac{\pi^2}{4}\right)^2\frac{\pi}{2}(1)$$
$$+ \left(\frac{\pi^2}{4}\right)^2\frac{\pi}{2}(0) = \frac{3}{32}\pi^5.$$

B. Here is another way to perform implicit differentiation (Chapter 2, Section 6):
If $y = f(x)$ is defined implicitly by the equation $F(x, y) = 0$, then
$$\frac{dy}{dx} = -\frac{F_x}{F_y}.$$
For example, if y is a function of x defined by $x^2 + y^2 = 1$, then letting
$F(x, y) = x^2 + y^2 - 1$,
$$y' = -\frac{F_x}{F_y} = -\frac{2x}{2y} = -\frac{x}{y}.$$
Likewise, if $z = f(x, y)$ is defined implicitly by $F(x, y, z) = 0$, then
$$\frac{\partial z}{\partial x} = -\frac{F_x}{F_z} \text{ and } \frac{\partial z}{\partial y} = -\frac{F_y}{F_z}.$$

3) Find y' where y is defined by
   $x^2y^3 + \cos(xy) = 0$.

Let $F(x, y) = x^2y^3 + \cos(xy)$.
$F_x = 2xy^3 - y\sin(xy)$
$F_y = 3x^2y^2 - x\sin(xy)$.
Thus $y' = -\frac{F_x}{F_y} = \frac{y\sin(xy) - 2xy^3}{3x^2y^2 - x\sin(xy)}$.

4) Find the equation of the tangent plane
   to $x^2 + 2y^2 + z^2 = 6$ at the point
   $(2, -1, 1)$.

Instead of solving for z and differentiating,
find $\frac{\partial z}{\partial x}$ and $\frac{\partial z}{\partial y}$ implicitly.
Let $F(x, y, z) = x^2 + 2y^2 + z^2 - 6$.
$$\frac{\partial z}{\partial x} = -\frac{F_x}{F_z} = -\frac{2x}{2z} = -\frac{x}{z}.$$
At $(2, -1, 1)$, $\frac{\partial z}{\partial x} = -2$.
$$\frac{\partial z}{\partial y} = -\frac{F_y}{F_z} = -\frac{4y}{2z} = -\frac{2y}{z}.$$
At $(2, -1, 1)$, $\frac{\partial z}{\partial y} = 2$.
The tangent plane is
$z - 1 = -2(x - 2) + 2(y + 1)$.

Directional Derivatives and the Gradient Vector

Concepts to Master

A. Directional derivative; Gradient; Maximum value of directional derivative
B. Tangent plane to a level surface

Summary and Focus Questions

A. The <u>directional derivative</u> of $f(x, y)$ at (x_0, y_0) in the direction of the unit vector $\mathbf{u} = \langle a, b \rangle$ is

$$D_\mathbf{u}f(x_0, y_0) = \lim_{h \to 0} \frac{f(x_0+ah, y_0+bh)-f(x_0, y_0)}{h} = f_x(x_0, y_0)a + f_y(x_0, y_0)b.$$

$D_\mathbf{u}f(x_0, y_0)$ may be interpreted as the slope of the tangent line to the graph of $z = f(x, y)$ in the $\mathbf{u}$ direction. It is also the instantaneous rate of change of z at (x_0, y_0) in the $\mathbf{u}$ direction.

The <u>gradient of $f(x, y)$</u> is $\nabla f(x, y) = f_x(x, y)\mathbf{i} + f_y(x, y)\mathbf{j}$.

Using the gradient, a directional derivative may be written
$$D_\mathbf{u}f(x_0, y_0) = \nabla f(x_0, y_0) \cdot \mathbf{u}. \qquad \text{(dot product of } \nabla f \text{ and } \mathbf{u}\text{)}$$

The direction in which $D_\mathbf{u}f(x_0, y_0)$ is maximum is
$$\mathbf{u} = \frac{\nabla f(x_0, y_0)}{|\nabla f(x_0, y_0)|}$$
with maximum value $|\nabla f(x_0, y_0)|$; that is, $z = f(x, y)$ increases most rapidly in the direction of the gradient. Likewise, $-\nabla f(x_0, y_0)$ is the direction in the xy-plane in which f decreases most rapidly.

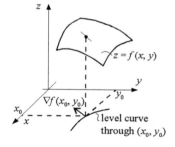

Directional derivatives and gradients for functions of three or more variables are defined similarly.

100

1) If $\mathbf{u} = \mathbf{i}$, then what is $D_\mathbf{u} f(x_0, y_0)$?

2) Find the directional derivative of

   a) $f(x, y) = x^2 y^2 - xy^3$ in the direction $\mathbf{u} = \frac{1}{2}\mathbf{i} - \frac{\sqrt{3}}{2}\mathbf{j}$.

   b) $f(x, y, z) = xy \sin z$ in the direction $\mathbf{u} = \left\langle \frac{1}{2}, -\frac{1}{\sqrt{2}}, \frac{1}{2} \right\rangle$ at the point P: $\left(2, 1, \frac{\pi}{6}\right)$.

3) Find the gradient of $f(x, y) = e^{xy}$.

4) Let $f(x, y, z) = xy^2 + yz^2$.

   a)   What is the maximum value of $D_\mathbf{u} f(1, 2, 1)$ as $\mathbf{u}$ varies?

   b)   What direction $\mathbf{u}$ gives the maximum value of $D_\mathbf{u} f(1, 2, 1)$?

   c)   Find $D_\mathbf{v} f(1, 2, 1)$ where $\mathbf{v} = \left\langle \frac{5}{6}, \frac{1}{2}, \frac{\sqrt{2}}{6} \right\rangle$.

The partial derivative $f_x(x_0, y_0)$.

$f_x = 2xy^2 - y^3$,
$f_y = 2x^2 y - 3xy^2$.
$D_\mathbf{u} f = \frac{1}{2}(2xy^2 - y^3)$
$\qquad\qquad - \frac{\sqrt{3}}{2}(2x^2 y - 3xy^2)$
$\qquad = \frac{2+3\sqrt{3}}{2} xy^2 - \frac{1}{2}y^3 - \sqrt{3}x^2 y$.

At P: $\left(2, 1, \frac{\pi}{6}\right)$,
$f_x = y \sin z = 1\left(\frac{1}{2}\right) = \frac{1}{2}$,
$f_y = x \sin z = 2\left(\frac{1}{2}\right) = 1$, and
$f_z = xy \cos z = 2(1) \cdot \frac{\sqrt{3}}{2} = \sqrt{3}$.
$D_\mathbf{u} f\left(2, 1, \frac{\pi}{6}\right) = \frac{1}{2}\left(\frac{1}{2}\right) + 1\left(-\frac{1}{\sqrt{2}}\right)$
$\qquad\qquad + \sqrt{3}\left(\frac{1}{2}\right)$
$\qquad = \frac{1 - 2\sqrt{2} + 2\sqrt{3}}{4}$.

$f_x = ye^{xy}$ and $f_y = xe^{xy}$, so
$\nabla f(x, y) = ye^{xy}\mathbf{i} + xe^{xy}\mathbf{j}$.

The maximum value is $|\nabla f(1, 2, 1)|$.
$\nabla f = \left\langle y^2, 2xy + z^2, 2yz \right\rangle$.
Thus $\nabla f(1, 2, 1) = \left\langle 4, 5, 4 \right\rangle$ and
$|\nabla f(1, 2, 1)| = \sqrt{4^2 + 5^2 + 4^2} = \sqrt{57}$.

$\mathbf{u}$ is the direction $\nabla f(1, 2, 1)$. But $\mathbf{u}$ must be a unit vector, so $\mathbf{u} = \left\langle \frac{4}{\sqrt{57}}, \frac{5}{\sqrt{57}}, \frac{4}{\sqrt{57}} \right\rangle$.

$D_\mathbf{v} f(1, 2, 1) = \nabla f(1, 2, 1) \cdot \mathbf{v}$
$\qquad = 4\left(\frac{5}{6}\right) + 5\left(\frac{1}{2}\right) + 4\left(\frac{\sqrt{2}}{6}\right) = \frac{35 + 4\sqrt{2}}{6}$.

B. Let S be a level surface $F(x, y, z) = 0$
and $P(x_0, y_0, z_0)$ be a point on S. The
gradient $\nabla F(x_0, y_0, z_0)$ is normal to the
surface S at the point P. Thus the tangent
plane to S at P has equation
$$\nabla F(x_0, y_0, z_0) \cdot \langle x - x_0, y - y_0, z - z_0 \rangle = 0,$$

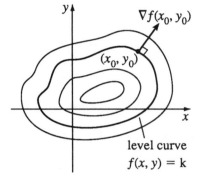

or

$$F_x(x_0, y_0, z_0)(x - x_0) + F_y(x_0, y_0, z_0)(y - y_0) + F_z(x_0, y_0, z_0)(z - z_0) = 0.$$

Likewise, the line normal to the surface at P_0 has equation
$$\frac{x - x_0}{F_x(x_0, y_0, z_0)} = \frac{y - y_0}{F_y(x_0, y_0, z_0)} = \frac{z - z_0}{F_z(x_0, y_0, z_0)}.$$

In the two dimensional case,
$z = f(x, y)$, the gradient of
$f(x, y)$ at (x_0, y_0) is a vector in the
xy-plane perpendicular to the level
curve of f at (x_0, y_0).

5) Find the equations of the tangent plane
and normal line to
$x^2 + 2y^2 - z^2 - 4xyz = 16$ at
$(1, 4, 1)$.

Let $F(x, y,$
$z) = x^2 + 2y^2 - z^2 - 4xyz - 16$
$\nabla F = \langle 2x - 4yz, 4y - 4xz, -2z - 4xy \rangle.$
$\nabla F(1, 4, 1) = \langle -14, 12, -18 \rangle.$
The tangent plane is
$-14(x - 1) + 12(y - 4) - 18(z - 1) = 0,$
$7x - 6y + 9z = 8.$
The normal line is
$\frac{x-1}{-14} = \frac{y-4}{12} = \frac{z-1}{-18}$, or $\frac{x-1}{7} = \frac{4-y}{6} = \frac{z-1}{9}.$

6) Given the level curve to $z = f(x, y)$ below, which vectors could represent gradients?

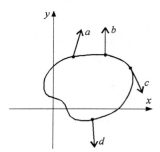

b and **d** appear to be normal to the level curve and could be gradients.

Maximum and Minimum Values

Concepts to Master

A. Local maximum; Local minimum; Critical point; Second Derivative Test
B. Absolute maximum and minimum; Absolute extrema on a closed, bounded set;
   Extreme value problems

Summary and Focus Questions

A. A function $z = f(x, y)$ has a local maximum at (a, b) if $f(a, b) \geq f(x, y)$ for all (x, y) in some open disk about (a, b). A local minimum at (a, b) is defined similarly but with $f(a, b) \leq f(x, y)$. These are local extrema of f.

A critical point (a, b) of f is a point in the domain of f for which both $f_x(a, b) = 0$ and $f_y(a, b) = 0$ or at least one of the partial derivatives does not exist.

If (a, b) is a local extremum for a continuous $z = f(x, y)$ then (a, b) is a critical point. The converse is false. A critical point that is not a local extremum is a saddle point.

The method of finding some critical points for $z = f(x, y)$ involves setting $f_x = 0$ and $f_y = 0$ and solving these two equations in two variables simultaneously. Often the system $f_x = 0$ and $f_y = 0$ is not linear and may be very difficult to solve. There are no general methods. One of the first approaches is to solve for one variable in terms of the other variable in one equation and substitute that value into the other equation.

Second Derivative Test: Suppose $z = f(x, y)$ has continuous second partial derivatives in a disk with center (a, b). Suppose $f_x(a, b) = 0$ and $f_y(a, b) = 0$ and let $D(a, b) = f_{xx}(a, b)f_{yy}(a, b) - [f_{xy}(a, b)]^2$.

Then the following holds:

Conditions	Conclusion
$D > 0$, $f_{xx} > 0$	f has a local minimum at $(a,\ b)$
$D > 0$, $f_{xx} < 0$	f has a local maximum at $(a,\ b)$
$D < 0$	f has a saddle point at $(a,\ b)$
$D = 0$	No conclusion is possible

A shorthand way to remember D is as a determinant:
$$D = \begin{vmatrix} f_{xx} & f_{xy} \\ f_{yx} & f_{yy} \end{vmatrix} = f_{xx}f_{yy} - (f_{xy})^2$$

1) Let $f(x,\ y) = y^3 - 24x - 3x^2y$.

   a)   Find the critical points of f.

Since f is a polynomial, critical points occur only where $f_x = 0$ and $f_y = 0$:
$$f_x = -24 - 6xy = 0$$
$$f_y = 3y^2 - 3x^2 = 0$$
To solve this, we observe from the second equation that $3x^2 = 3y^2$, so $x = y$ or $x = -y$. Substituting $x = y$ in the first:
$$-24 - 6x(x) = 0$$
$$-24 - 6x^2 = 0$$
$$6x^2 = -24$$
This has no solution.
Substituting $x = -y$ in the first:
$$-24 - 6x(-x) = 0$$
$$-24 + 6x^2 = 0$$
$$6x^2 = 24$$
$$x^2 = 4,\ x = 2,\ -2$$
When $x = 2$, $y = -2$ and when $x = -2$, $y = 2$. There are no points where f_x or f_y does not exist. Thus the critical points are $(2,\ -2)$ and $(-2,\ 2)$.

b) Find the local extrema of $f(x, y)$.

The extrema are among the critical points.
$f_{xx} = -6y$, $f_{xy} = -6x$, $f_{yy} = 6y$.
$D = (-6y)(6y) - (-6x)^2$
$\quad = -36y^2 - 36x^2$
At both $(2, -2)$ and $(-2, 2)$,
$D = -288 < 0$.
Thus both $(2, -2)$ and $(-2, 2)$ are saddle points. There are no local extrema.

2) Suppose f has continuous second derivatives and has four critical points with the following information about the second derivatives.

Point	f_{xx}	f_{yy}	f_{xy}
$A: (7, 1)$	8	2	-4
$B: (1, 2)$	1	9	2
$C: (0, 4)$	-6	3	1
$E: (-1, 3)$	-2	-5	3

Classify each critical point.

We calculate $D = f_{xx}f_{yy} - (f_{xy})^2$ for each:

Point	D
A	$8(2) - (-4)^2 = 0$
B	$1(9) - 2^2 = 5$
C	$-6(3) - 1^2 = -19$
E	$-2(-5) - 3^2 = 1$

By the Second Derivative Test:
B is a local minimum, C is a saddle point, E is a local maximum.
We cannot conclude anything about A from the information given.

3) a) Find all critical points of
$$f(x, y) = x^2 - 4xy + 2x - \tfrac{4}{15}(5y + 41)^{3/2}$$

$f_x = 2x - 4y + 2 = 2(x - 2y + 1).$
$f_y = -4x - \tfrac{4}{15}\left(\tfrac{3}{2}\right)(5y + 41)^{1/2}(5)$
$\quad = -2(2x + \sqrt{5y + 41}).$

Set both f_x and f_y to 0.

$x - 2y + 1 = 0 \qquad 2x + \sqrt{5y + 41} = 0.$

Solve for x in the first:

$x = 2y - 1$

and substitute:

$2(2y - 1) + \sqrt{5y + 41} = 0$
$2 - 4y = \sqrt{5y + 41}$
$4 - 16y + 16y^2 = 5y + 41$
$16y^2 - 21y - 37 = 0$
$(16y - 37)(y + 1) = 0$
$y = \tfrac{37}{16}, \; y = -1.$

From $x = 2y - 1$, the possible critical points are $\left(\tfrac{29}{8}, \tfrac{37}{16}\right)$ and $(-3, -1)$.
$f_x(-3, -1) = f_y(-3, -1) = 0$ but $f_y\left(\tfrac{29}{8}, \tfrac{37}{16}\right) \neq 0$ so $(-3, -1)$ is the only critical point.

b) Classify the critical point in part a)

$f_{xx} = 2, \; f_{xy} = -4$ and
$f_{yy} = \tfrac{1}{2}(5y + 41)^{-1/2}(5) = \tfrac{5}{2\sqrt{5y+41}}.$

At $(-3, -1)$, $f_{yy} = \tfrac{5}{12}.$

$f_{xx}f_{yy} - (f_{xy})^2 = 2(-4) - \left(\tfrac{5}{12}\right)^2$ which is negative. Thus $(-3, -1)$ is a saddle point.

B. The <u>absolute maximum</u> of f is the value $f(a, b)$ for some (a, b) such that $f(a, b) \geq f(x, y)$ for all (x, y) in the domain of f. <u>Absolute minimum</u> is defined as above but with $f(a, b) \leq f(x, y)$. These are the <u>extreme values</u> of f.

A continuous function whose domain is closed (contains all its boundary points) and bounded has an absolute maximum and absolute minimum.

To find the absolute extrema of continuous $f(x, y)$ on a closed and bounded set D, compute $f(x, y)$ at critical points in D and along the boundary of D.

4)   Does the absolute extrema exist for
     $f(x, y) = x^2 + y^2$ with domain
     $D = \{(x, y)|0 \leq x \leq 2, 0 \leq y \leq 3\}$?

5)   Find the extreme values on
     $f(x, y) = 10 - x^2 - 2y^2 + 2x + 8y$
     with domain D as graphed:

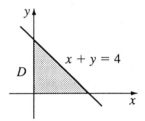

Yes, f is continuous and D is both closed
and bounded.

First find the critical points
$f_x = -2x + 2 = 0$, at $x = 1$.
$f_y = -4y + 8 = 0$, at $y = 2$.
$(1, 2)$ is the only critical point and it is
within D. D has three boundary lines:
l_1(x-axis): $y = 0$, $0 \leq x \leq 4$.
Here $f(x, y) = f(x, 0) = 10 - x^2 + 2x$
has a maximum at $(1, 0)$ and a minimum at
$(4, 0)$.
l_2(y-axis): $x = 0$, $0 \leq y \leq 4$.
Here $f(x, y) = f(0, y) = 10 - 2y^2 + 8y$
has a maximum at $(0, 2)$ and a minimum at
$(0, 4)$.
l_3($x + y = 4$): $y = 4 - x$, $0 \leq x \leq 4$.
Here $f(x, y) = f(x, 4 - x)$
$= 10 - x^2 - (4 - x)^2 + 2x + 8(4 - x)$
$= 26 - 2x^2 + 6x$
has a maximum at $x = \frac{3}{2}$, $y = \frac{5}{2}$ and a
minimum at $(4, 0)$.
We compute the corresponding $f(x, y)$
values and summarize in a table:

(x, y)	How found?	$f(x, y)$
$(1, 0)$	max on l_1	11
$(4, 0)$	min on both l_1 and l_3	2
$(0, 2)$	max on l_2	18
$(0, 4)$	min on l_2	10
$\left(\frac{3}{5}, \frac{5}{2}\right)$	max on l_3	18.25
$(1, 2)$	critical point	19

The absolute maximum is 19 and occurs at
$(1, 2)$. The absolute minimum is 2 and
occurs at $(4, 0)$.

6)   Show that among all rectangular
     parallelepipeds with volume 1 cubic
     inch, the one with smallest surface
     area is a cube.

We are given $xyz = 1$ and must minimize
$S = 2xy + 2xz + 2yz$.
From $xyz = 1$, $z = \frac{1}{xy}$.

Thus $S = 2xy + 2x\left(\frac{1}{xy}\right) + 2y\left(\frac{1}{xy}\right)$
$\qquad = 2xy + \frac{2}{y} + \frac{2}{x}$.
$S_x = 2y - \frac{2}{x^2} = 0$
$\qquad 2y = \frac{2}{x^2}, \; y = \frac{1}{x^2}$.
$S_y = 2x - \frac{2}{y^2} = 0$
$\qquad 2x = \frac{2}{y^2}, \; x = \frac{1}{y^2}$.

From $y = \frac{1}{x^2}$ and $x = \frac{1}{y^2}$, $y = y^4$.
Thus $y = 0$ or $y = 1$.
$y = 0$ is not possible so $y = 1$.
Thus $x = \frac{1}{y^2} = 1$ and $z = \frac{1}{xy} = \frac{1}{1 \cdot 1} = 1$.
Therefore the object is a cube with 1-inch
edges.

Lagrange Multipliers

Concepts to Master

Solution to extreme value problems using Lagrange Multipliers

Summary and Focus Questions

If $f(x, y)$ and $g(x, y)$ have continuous partial derivatives and (a, b) is a local extremum for f when restricted to $g(x, y) = k$ (a constraint), then there is a number λ called a <u>Lagrange multiplier</u> such that
$$\nabla f(a, b) = \lambda \nabla g(a, b).$$

Thus solving the constrained extremum problem above is the same as solving the equations $\nabla f = \lambda \nabla g$ and $g(x, y) = k$.

The solution to the constrained optimum problem involving three variables is determined by solving
$$\nabla f(x, y, z) = \lambda \nabla g(x, y, z), \quad g(x, y, z) = k.$$

For problems with two constraints $g_1(x, y, z) = k_1$, $g_2(x, y, z) = k_2$:
$$\nabla f(x, y, z) = \lambda_1 \nabla g(x, y, z) + \lambda_2 \nabla g(x, y, z).$$

1) Find the extrema of
$f(x, y) = x^2 - 4xy + 2y^2$ subject to
$2x - 3y = 14$.

$\nabla f = \langle 2x - 4y, -4x + 4y \rangle$.
Let $g(x, y) = 2x - 3y$.
Then $\nabla g = \langle 2, -3 \rangle$.
From $\nabla f = \lambda \nabla g$, $g(x, y) = 14$, we have three equations with variables x, y, λ.
(1) $2x - 4y = 2\lambda$
(2) $-4x + 4y = -3\lambda$
(3) $2x - 3y = 14$.
Add (1) and (2) to get $-2x = -\lambda$, $x = \frac{1}{2}\lambda$
Multiply (1) by 2 and add (2):
$-4y = \lambda$, $y = -\frac{1}{4}\lambda$.

From (3), $2\left(\frac{1}{2}\lambda\right) - 3\left(-\frac{1}{4}\lambda\right) = 14$,
$\frac{7}{4}\lambda = 14$, $\lambda = 8$.
Thus $x = \frac{1}{2}\lambda = 4$, $y = -\frac{1}{4}\lambda = -2$.
$f(4, -2) = 56$. By inspection, 56 is the
absolute minimum of $x^2 - 4xy + 2y^2$
subject to $2x - 3y = 14$.

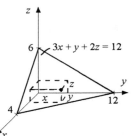

2) Find the system of equations for solving:
   Maximize
   $f(x, y, z) = x^2 + 6xy^2 + 3xz^2 + yz^3$
   subject to $x + y + z^2 = 2$ and
   $x^2 + y + z = 3$.

$\nabla f = $
$\langle 2x + 6y^2 + 3z^2,\ 12xy + z^3,\ 6xz + 3yz^2 \rangle$.
Let $g_1(x, y, z) = x + y + z^2$,
$\nabla g_1 = \langle 1,\ 1,\ 2z \rangle$ and
$g_2(x, y, z) = x^2 + y + z$,
$\nabla g_2 = \langle 2x,\ 1,\ 1 \rangle$.
The system of five equations with
variables x, y, z, λ_1 and λ_2 from
$\nabla f = \lambda_1 \nabla g_1 + \lambda_2 \nabla g_2$,
$g_1(x, y, z) = 2$, $g_2(x, y, z) = 3$ is
$$2x + 6y^2 + 3z^2 = \lambda_1 + 2x\lambda_2$$
$$12xy + z^3 = \lambda_1 + \lambda_2$$
$$6xz + 3yz^2 = 2z\lambda_1 + \lambda_2$$
$$x + y + z^2 = 2$$
$$x^2 + y + z = 3.$$

3) Find the volume of the largest rectangular box in the first octant with three faces in the coordinate planes and one vertex in the plane
   $3x + y + 2z = 12$.

We need to find the maximum volume of
the box, $V = xyz$, subject to
$3x + y + 2z = 12$.

Let $g(x, y, z) = 3x + y + 2z$.
$\nabla V(x, y, z) = \langle yz, xz, xy \rangle$ and
$\nabla g(x, y, z) = \langle 3, 1, 2 \rangle$.
Thus $\nabla f = \lambda \nabla g$ becomes
$$yz = 3\lambda$$
$$xz = \lambda$$
$$xy = 2\lambda$$
Multiplying by x, y, z respectively:
$$xyz = 3x\lambda$$
$$xyz = y\lambda$$
$$xyz = 2z\lambda$$
Therefore $3x\lambda = y\lambda = 2z\lambda$ and for $\lambda \neq 0$
$$3x = y = 2z.$$
Then $3x + y + 2z = 12$ becomes
$y + y + y = 12$, so $y = 4$.
Thus $3x = 4$ or $x = \frac{4}{3}$ and $2z = 4$ or
$z = 2$. The maximum volume is a box
$\frac{4}{3} \times 4 \times 2$.

13

Multiple Integrals

"SORRY I'M LATE — I WAS WORKING OUT PI TO 5,000 PLACES."

Cartoons courtesy of Sidney Harris. Used by permission.

Double Integrals over Rectangles

Concepts to Master

A. Double Riemann sum; Double integral; Integrable
B. Interpretation of the double integral as a volume

Summary and Focus Questions

A. A <u>partition</u> P of the rectangular region $R = [a, b] \times [c, d]$ is formed by partitioning $[a, b]$ and $[c, d]$ and drawing parallel lines through the partitioning points as in the figure.

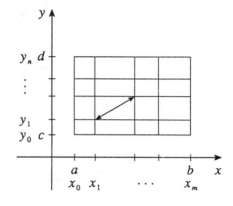

The i, j-th rectangle R_{ij} has area $\Delta A_{ij} = \Delta x_i \Delta y_i$. The <u>norm</u> of P, $\|P\|$, is the largest length of the diagonals of all R_{ij}.

For a given partition a double <u>Riemann sum</u> for a function f is a number of the form

$$\sum_{i=1}^{m} \sum_{j=1}^{n} f(x_i^*, y_j^*) \Delta A_{ij}$$

where (x_i^*, y_j^*) is a point chosen from R_{ij}.

Just as with single integrals, <u>the double integral of f over R</u> is a limit of Riemann sums:

$$\iint_R f(x, y)\,dA = \iint_R f(x, y)\,dx\,dy = \lim_{\|P\| \to 0} \sum_{i=1}^{m} \sum_{j=1}^{n} f(x_i^*, y_j^*) \Delta A_{ij}$$

If the limit exists f is <u>integrable</u> over R.

1) Find the Riemann sum for
   $f(x, y) = x^2 + y^2$ over
   $R = [1, 9] \times [-1, 3]$ with partition
   determined by $x_0 = 1$, $x_1 = 3$,
   $x_2 = 5$, $x_3 = 9$, and $y_0 = -1$, $y_1 = 1$,
   $y_2 = 3$. Use midpoints of each sub-
   rectangle for (x_i^*, y_j^*).

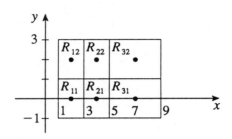

	R_{11}	R_{12}	R_{21}	R_{22}	R_{31}	R_{32}
ΔA_{ij}	4	4	4	4	8	8
x_i^*	2	2	4	4	7	7
y_j^*	0	2	0	2	0	2
$f(x_i^*, y_i^*)$	4	8	16	20	49	53

The Riemann sum is

$$\sum_{i=1}^{3} \sum_{j=1}^{2} f(x_i^*, y_i^*) \Delta A_{ij}$$

$$= 4(4) + 8(4) + 16(4) + 20(4)$$
$$+ 49(8) + 53(8)$$
$$= 1008$$

B. Let $R = [a, b] \times [c, d]$.
   For $f(x, y) \geq 0$ the double integral
   $\iint_R f(x, y) dA$ may be interpreted as
   the volume of the solid above R and
   under the surface $z = f(x, y)$.

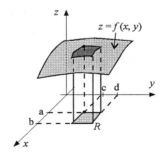

2) Write a double integral expression for
    the volume of the pictured solid.

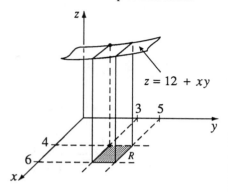

$z = 12 + xy$

3) Let $R = [1, 4] \times [5, 9]$.
    Evaluate $\iint\limits_R 2\,dA$.

Let R be the rectangular region $4 \leq x \leq 6$,
$3 \leq y \leq 5$. The volume pictured is
$\iint\limits_R (12 + xy)\,dA$.

$\iint\limits_R 2\,dA$ is the volume of the solid under
$f(x, y) = 2$. The solid is a rectangular
block with a 3×4 base and height 2.
$\iint\limits_R 2\,dA = 3(4)2 = 24$.

Iterated Integrals

Concepts to Master

Iterated integrals; Double integrals written as iterated integrals; Fubini's Theorem

Summary and Focus Questions

An <u>iterated integral</u> is defined as
$$\int_a^b \int_c^d f(x,\ y)dy\ dx = \int_a^b \left[\int_c^d f(x,\ y)dy \right] dx.$$
First perform partial integration of $f(x,\ y)$ with respect to y and then integrate the result with respect to x.

<u>Fubini's Theorem</u>, stated below, allows us to write a double integral as an iterated integral:

When R is a closed rectangle $a \leq x \leq b$, $c \leq y \leq d$,

$$\iint\limits_R f(x,\ y)dA = \int_a^b \int_c^d f(x,\ y)dy\ dx = \int_c^d \int_a^b f(x,\ y)dx\ dy.$$

1) Evaluate

a) $\int_1^3 \int_0^2 (x^2 + 4y)dy\ dx$

$\int_1^3 \left[\int_0^2 (x^2 + 4y)dy \right] dx$

$= \int_1^3 \left[(x^2 y + 2y^2) \Big|_{y=0}^{2} \right] dx$

$= \int_1^3 (2x^2 + 8)dx$

$= \left(\frac{2}{3}x^3 + 8x \right) \Big|_{x=1}^{3} = \frac{100}{3}.$

b) $\int_0^2 \int_0^3 8xy \ dx \ dy$

$\int_0^2 \left[\int_0^3 8xy \ dx \right] dy$

$\quad = \int_0^2 \left(4x^2 y \Big|_{x=0}^{3} \right) dy$

$\quad = \int_0^2 36y \ dy = 18y^2 \Big|_{y=0}^{2}$

$\quad = 72.$

c) $\int_0^3 \int_0^2 8xy \ dy \ dx$

72. This is the iterated integral in question 1 b), written in reverse order.

2) Find $\iint\limits_R (xy + e^x) dA$
    where $R =$
    $\{(x, y) | 0 \le x \le 1, 0 \le y \le 2\}$.

$\iint\limits_R (xy + e^x) dA = \int_0^2 \int_0^1 (xy + e^x) dx \ dy$

$\quad = \int_0^2 \left(\frac{x^2 y}{2} + e^x \right) \Big|_{x=0}^{1} dy$

$\quad = \int_0^2 \left(\frac{y}{2} + e - 1 \right) dy$

$\quad = \left(\frac{y^2}{4} + (e - 1)y \right) \Big|_{y=0}^{2}$

$\quad = 2e - 1.$

3) Write an iterated integral for
    $\iint\limits_R (x^2 + y^2) dA$ where R is the
    rectangle $0 \le x \le 4, 3 \le y \le 5$.

$\int_0^4 \int_3^5 (x^2 + y^2) dy \ dx$

$\left(\text{or } \int_3^5 \int_0^4 (x^2 + y^2) dx \ dy \right).$

Double Integrals over General Regions

Concepts to Master

A. Evaluate a double integral over a general region as an iterated integral
B. Change the order of integration of an iterated integral
C. Properties of double integrals

Summary and Focus Questions

A. Let D be a region in the plane and $f(x, y)$ be continuous on D. If D may be described in either of the following ways, then $\iint\limits_{D} f(x, y)dA$ may be written as an iterated integral:

Type I

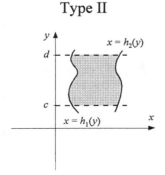

D described as
$a \le x \le b,\ g_1(x) \le y \le g_2(x).$

$$\iint\limits_{D} f(x, y)dA$$
$$= \int_a^b \int_{g_1(x)}^{g_2(x)} f(x, y)dy\ dx$$

Type II

D described as
$c \le y \le d,\ h_1(y) \le x \le h_2(y).$

$$\iint\limits_{D} f(x, y)dA$$
$$= \int_c^d \int_{h_1(y)}^{h_2(y)} f(x, y)dx\ dy$$

1)   Write each as an iterated integral:

a) $\iint_D xy^2 \, dA$, where D is the region below.

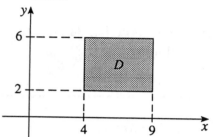

$\int_2^6 \int_4^9 xy^2 \, dx \, dy$ or $\int_4^9 \int_2^6 xy^2 \, dy \, dx$.

b) $\iint_D e^{xy} \, dA$, where D is the region below.

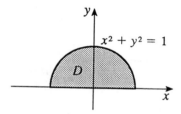

The region may be described as Type I:

$-1 \le x \le 1,\ -\sqrt{1-x^2} \le y \le \sqrt{1-x^2}$.

The integral is $\int_{-1}^1 \int_{-\sqrt{1-x^2}}^{\sqrt{1-x^2}} e^{xy} \, dy \, dx$.

c) $\iint_D (x+y)dA$, where D is the region below.

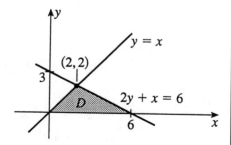

Describing D as Type I would be difficult (the "top" function would be defined piecewise). D is easier to describe as Type II. $[h_1(y) = y$ and from $2y + x = 6$, $h_2(y) = 6 - 2y]$. The double integral is $\int_0^2 \left[\int_y^{6-2y} (x+y)dx \right] dy$.

2) Evaluate $\int_1^2 \int_y^{y^2} 4xy \ dx \ dy$.

$$\int_1^2 \int_y^{y^2} 4xy \ dx \ dy = \int_1^2 2x^2 y \Big|_{x=y}^{y^2} \ dy$$
$$= \int_1^2 [2(y^2)^2 y - 2y^2 y] dy$$
$$= \int_1^2 (2y^5 - 2y^3) dy$$
$$= \left(\frac{y^6}{3} - \frac{y^4}{2} \right) \Big|_{y=1}^{2}$$
$$= \left(\frac{64}{3} - \frac{16}{2} \right) - \left(\frac{1}{3} - \frac{1}{2} \right) = \frac{27}{2}.$$

B. An iterated integral corresponds to a double integral over a region R. If R may be described as both Type I and Type II it is possible to write the given form of the iterated integral in reverse form.

For example, $\int_0^3 \int_0^{6-2x} (x+y) dy \ dx$ corresponds to the region
$R: 0 \le x \le 3, \ 0 \le y \le 6 - 2x$.
From $y = 6 - 2x$, $x = 3 - \frac{y}{2}$.
Hence R may be described
$0 \le y \le 6, \ 0 \le x \le 3 - \frac{y}{2}$. Thus
$\int_0^3 \int_0^{6-2x} (x+y) dy \ dx$
$= \int_0^6 \int_0^{3-y/2} (x+y) dx \ dy$.

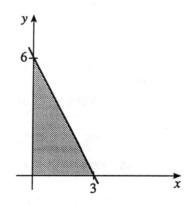

3) Change the order of integration of
$\int_0^4 \int_0^{\sqrt{4-y^2}} xy \ dy \ dx$.

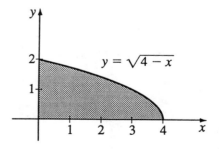

The iterated integral is $\iint_D xy \, dA$ where D is shown above. D may be rewritten as a Type II region:
$$0 \le y \le 2$$
$$0 \le x \le 4 - y^2$$
Thus the double integral is
$$\int_0^2 \int_0^{4-y^2} xy \, dx \, dy.$$

C. If a region D may be described as the union of two nonoverlapping regions D_1 and D_2 then
$$\iint_D f(x, y) dA$$
$$= \iint_{D_1} f(x, y) dA + \iint_{D_2} f(x, y) dA$$

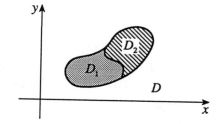

In particular, if D is neither Type I nor Type II it may be possible to partition D into subregions of these types. The double integral of D is the sum of the double integrals over the subregions.

The area of a region D in the plane may be written as $\iint_D dA$, which may be evaluated using iterated integrals.

If f is integrable over D and $m \le f(x, y) \le M$ for all $(x, y) \in D$, then $mK \le \iint_D f(x, y) dA \le MK$, where K is the area of D.

4) Write an iterated integral for $\iint_D x^2 y \, dA$ where the region D is bounded by $y = 5x$, $y = x$, $x + y = 6$.

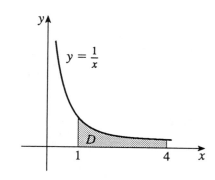

D may be divided into two Type II regions by the line $x = 1$. The area is
$\int_0^1 \int_x^{5x} x^2 y \, dy \, dx + \int_1^3 \int_x^{6-x} x^2 y \, dy \, dx$.

5) Write an iterated integral for the shaded area.

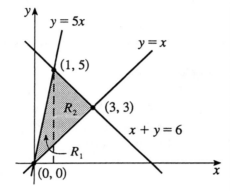

6) True or False:
$\int_0^1 \int_3^5 x^2 y \, dy \, dx$
$= \int_0^1 \int_3^4 x^2 y \, dy \, dx$
$+ \int_0^1 \int_4^5 x^2 y \, dy \, dx$.

7) Let $D =$
$\{(x, y)|1 \le x \le y^2, 1 \le y \le 2\}$.
Using $1 \le x^2 y \le 32$ find bounds on
$\iint\limits_D x^2 y \, dA$

D may be described as $1 \le x \le 4$,

$0 \le y \le \frac{1}{x}$. The area is $\int_1^4 \int_0^{1/x} dy \, dx$.

True.

$m = 1$ and $M = 32$.
The area of D is
$$\iint\limits_D dx \, dy = \int_1^2 \int_1^{y^2} dx \, dy$$
$$= \int_1^2 1 \Big|_{x=1}^{y^2} dy$$
$$= \int_1^2 (y^2 - 1) dy = \left(\frac{y^3}{3} - y\right)\Big|_{y=1}^2$$
$$= \frac{4}{3}.$$
Thus
$1\left(\frac{4}{3}\right) \le \iint\limits_D x^2 y \, dA \le 32\left(\frac{4}{3}\right)$
or $\frac{4}{3} \le \iint\limits_D x^2 y \, dA \le \frac{128}{3}$.

Double Integrals in Polar Coordinates

Concepts to Master

Rewrite an iterated integral in rectangular coordinates as an iterated integral in polar coordinates

Summary and Focus Questions

Let D be a region in the plane and $f(x, y)$ be continuous on D. If D may be described using polar coordinates in either of the following ways, then $\iint_D f(x, y)\,dA$ may be written as an iterated integral in polar coordinates:

Type I

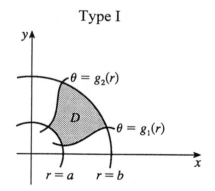

D described as
$a \leq r \leq b,\ g_1(r) \leq \theta \leq g_2(r).$

$$\iint_D f(x, y)\,dA$$
$$= \int_a^b \int_{g_1(r)}^{g_2(r)} f(r\cos\theta,\ r\sin\theta)r\ d\theta\ dr$$

Type II

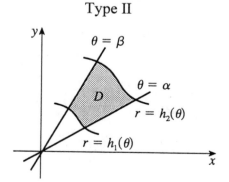

D described as
$\alpha \leq \theta \leq \beta,\ h_1(\theta) \leq r \leq h_2(\theta).$

$$\iint_D f(x, y)\,dA$$
$$= \int_\alpha^\beta \int_{h_1(\theta)}^{h_2(\theta)} f(r\cos\theta,\ r\sin\theta)r\ dr\ d\theta$$

1) Rewrite, using polar coordinates:

a) $\iint\limits_D x \; dA$ where D is the region below.

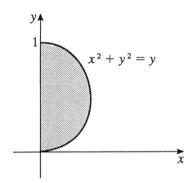

$x^2 + y^2 = y$

Converting $x^2 + y^2 = y$ to a polar equation gives $r^2 = r \sin \theta$, so $r = \sin \theta$.
The region D is Type II:
$0 \le \theta \le \frac{\pi}{2}, 0 \le r \le \sin \theta$.
Therefore the double integral is
$\int_0^{\pi/2} \int_0^{\sin \theta} (r \cos \theta) r \; dr \; d\theta$
(remember the "extra" r factor)
$= \int_0^{\pi/2} \int_0^{\sin \theta} r^2 \cos \theta \; dr \; d\theta.$

b) $\int_0^4 \int_0^x xy \; dy \; dx$

The corresponding region D is given by
$0 \le x \le 4, 0 \le y \le x$.

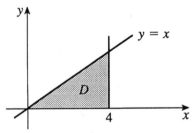

$y = x$

D

4

D may be described as a Type II region in polar coordinates:
$0 \le \theta \le \frac{\pi}{4}, 0 \le r \le 4 \sec \theta$
(The line $x = 4$ is $r \cos \theta = 4$, or $r = 4 \sec \theta$.)
Thus the double integral is
$\int_0^{\pi/4} \int_0^{4 \sec \theta} (r \cos \theta)(r \sin \theta) r \; dr \; d\theta$

$= \int_0^{\pi/4} \int_0^{4 \sec \theta} \frac{r^3 \sin 2\theta}{2} \; dr \; d\theta.$

2) Write a polar iterated integral for the volume under $z = x^2 + y^2$ and above the region given here.

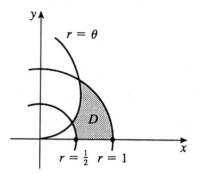

The region is Type I:
$\frac{1}{2} \le r \le 1, 0 \le \theta \le r$.
The volume under $z = x^2 + y^2$ is
$\iint\limits_{D} (x^2 + y^2) dA = \int_{1/2}^{1}\int_{0}^{r} r^2 r \; d\theta \; dr$
$\qquad = \int_{1/2}^{1}\int_{0}^{r} r^3 \; d\theta \; dr$.

Section 13.5

Applications of Double Integrals

Concepts to Master

Mass of a lamina; Center of mass; Moments of inertia

Summary and Focus Questions

Suppose a lamina (a flat plane area representing a distribution of matter) is described as a region D with density $\rho(x, y)$ at each point in D. Let ρ be continuous.

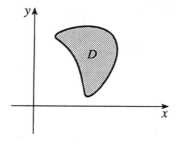

The <u>mass of the lamina</u> is
$$m = \iint_D \rho(x, y)dA.$$

<u>Moments:</u>

Moment of mass with respect to the x-axis is $M_x = \iint_D y\rho(x, y)dA$.

Moment of mass with respect to the y-axis is $M_y = \iint_D x\rho(x, y)dA$.

The center of mass is $(\overline{x}, \overline{y})$, where $\overline{x} = \frac{M_y}{m}$ and $\overline{y} = \frac{M_x}{m}$.

<u>Moments of Inertia:</u>

Moment of inertia about the x-axis is $I_x = \iint_D y^2\rho(x, y)dA$.

Moment of inertia about the x-axis is $I_y = \iint_D x^2\rho(x, y)dA$.

Moment of inertia about the origin is $I_0 = I_x + I_y$.

1) Find the mass of the lamina below whose density at each point (x, y) is $72xy$.

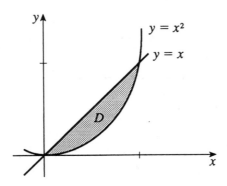

$$m = \iint_D 72xy \ dA = \int_0^1 \int_{x^2}^x 72xy \ dy \ dx$$
$$= \int_0^1 36xy^2 \Big|_{y=x^2}^{x} dx$$
$$= \int_0^1 (36x^3 - 36x^5) dx$$
$$= (9x^4 - 6x^6) \Big|_{x=0}^{1} = 3.$$

2) Find the y-coordinate of the center of mass of the lamina in question 1.

$$M_x = \iint_D y(72xy) dA = \iint_D 72xy^2 \ dA$$
$$= \int_0^1 \int_{x^2}^x 72xy^2 \ dy \ dx$$
$$= \int_0^1 (24x^4 - 24x^7) dx = \tfrac{9}{5}.$$

Thus $\bar{y} = \frac{M_x}{m} = \frac{9/5}{3} = \frac{3}{5}.$

3) Find an iterated integral for the moment of inertia about the y-axis of the lamina in question 1.

$$I_y = \iint_D x^2(72xy) dA$$
$$= \int_0^1 \int_{x^2}^x 72x^3 y \ dy \ dx.$$

Surface Area

Concepts to Master

Area of a surface given by the equation $z = f(x, y)$

Summary and Focus Questions

The area of the surface $z = f(x, y)$ where f has continuous partial derivatives over a region D is

$$\iint_D \sqrt{[f_x(x, y)]^2 + [f_y(x, y)]^2 + 1} \ dA.$$

1) Write an iterated integral for the surface area of the paraboloid $z = x^2 + 2y^2$ above the triangular region D:

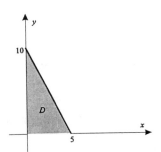

$f_x = 2x, \ f_y = 4y.$
The region D is $0 \le x \le 5$,
$0 \le y \le 10 - 2x$.
The surface area is
$$\iint_D \sqrt{(2x)^2 + (4y)^2 + 1} \ dA$$
$$= \int_0^5 \int_0^{10-2x} \sqrt{4x^2 + 16y^2 + 1} \ dy \ dx.$$

Triple Integrals

Concepts to Master

A. Triple Integrals; Fubini's Theorem (for writing triple integrals as iterated integrals)

B. Applications of the triple integral as volume, mass, center of mass, and moment of inertia

Summary and Focus Questions

A. Every concept for double integrals is now extended to three dimensions. A triple integral of a function $f(x, y, z)$ over a region E in a three-dimensional space $\iiint_E f(x, y, z)dV$ is defined in the usual way as a limit of Riemann sums:

$$\lim_{\|P\| \to 0} \sum_{i=1}^{l} \sum_{j=1}^{m} \sum_{k=1}^{n} f(x_i^*, y_j^*, z_k^*)\Delta V_{ijk}$$

where P is a partition of E into sub-boxes and (x_i^*, y_j^*, z_k^*) is chosen from the i-j-k-th sub-box whose volume is ΔV_{ijk}.

Fubini's Theorem: A triple integral may be evaluated as one of six possible iterated integrals. For example, if E can be described by $a \leq x \leq b$, $g_1(x) \leq y \leq g_2(x)$, $h_1(x, y) \leq z \leq h_2(x, y)$ then

$$\iiint_E f(x, y, z)dV = \int_a^b \int_{g_1(x)}^{g_2(x)} \int_{h_1(x,y)}^{h_2(x,y)} f(x, y, z)dz \, dy \, dx.$$

As with iterated double integrals, these are evaluated "from the inside out," that is, integrate with respect to z and substitute the limits of integration, then with respect to y and substitute the limits, and finally with respect to x and substitute the limits.

1) Evaluate $\int_1^2 \int_1^y \int_0^{x+y} 12x \; dz \; dx \; dy$.

$\int_1^2 \int_1^y \int_0^{x+y} 12x \; dz \; dx \; dy$

$\int_1^2 \int_1^y \left(12xz \Big|_{z=0}^{x+y} \right) dx \; dy$

$\int_1^2 \int_1^y (12x^2 + 12xy) dx \; dy$

$\int_1^2 \left((4x^3 + 6x^2 y) \Big|_{x=1}^{y} \right) dy$

$\int_1^2 (4y^3 + 6y^3) - (4 + 6y) dy$

$\int_1^2 (10y^3 - 6y - 4) dy$

$\qquad = \left(\frac{10}{4} y^4 - 3y^2 - 4y \right) \Big|_{y=1}^{2} = \frac{9}{2}.$

2) $\int_1^4 \int_2^5 \int_3^7 xyz \; dx \; dz \; dy$
$\qquad = \int_?^? \int_?^? \int_?^? xyz \; dz \; dy \; dx$

$\int_3^7 \int_1^4 \int_2^5 xyz \; dz \; dy \; dx.$

3) Write an iterated integral for $\iiint\limits_E x \; dV$, where E is the region cut off in the first octant by the plane $x + 2y + 4z = 8$.

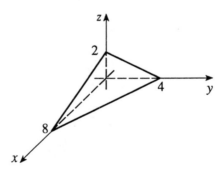

There are six possible iterated integrals as answers, each depending upon how the solid E is represented. We write E as $0 \le z \le 2, \; 0 \le y \le 4 - 2z,$ $0 \le x \le 8 - 2y - 4z.$
Thus
$\iiint\limits_E x \; dV = \int_0^2 \int_0^{4-2z} \int_0^{8-2y-4z} x \; dx \; dy \; dz.$

4) Write $\iiint\limits_{E}(y+z)dV$ as an iterated integral where E is the solid ball of radius 2 about the origin.

The region is inside the sphere $x^2 + y^2 + z^2 = 4$. E may be described by $-2 \le x \le 2$,

$$-\sqrt{4-x^2} \le y \le \sqrt{4-x^2},$$

$$-\sqrt{4-x^2-y^2} \le z \le \sqrt{4-x^2-y^2}.$$

Thus $\iiint\limits_{E}(y+z)dV =$

$$\int_{-2}^{2}\int_{-\sqrt{4-x^2}}^{\sqrt{4-x^2}}\int_{-\sqrt{4-x^2-y^2}}^{\sqrt{4-x^2-y^2}}(y+z)dz\ dy\ dx.$$

5) Write $\int_0^1\int_0^y\int_0^{1-y} y^2\ dx\ dz\ dy$ as an iterated integral in the form $\iiint y^2\ dz\ dy\ dx$.

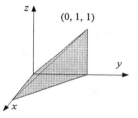

(0, 1, 1)

The given iterated integral represents the region above, which may be written as $0 \le y \le 1, 0 \le z \le y, 0 \le x \le 1-y$. For $\iiint y^2\ dz\ dy\ dx$, the region is $0 \le x \le 1, 0 \le y \le 1-x, 0 \le z \le y$. Thus the integral is $\int_0^1\int_0^{1-x}\int_0^y y^2\ dz\ dy\ dx$.

B. The volume of a solid E is $\iiint\limits_{E} dV$. If a solid E has a continuous density $\rho(x,\ y,\ z)$, the <u>mass</u>, m, is $\iiint\limits_{E}\rho(x,\ y,\ z)dV$.

The three <u>moments</u> are:

about xy-plane	about xz-plane	about yz-plane
$M_{xy} = \iiint\limits_{E} z\rho(x,\ y,\ z)dV$	$M_{xz} = \iiint y\rho(x,\ y,\ z)dV$	$M_{yz} = \iiint\limits_{E} x\rho(x,\ y,\ z)dV$

The <u>moments of inertia</u> about the axes are:

$$I_x = \qquad\qquad I_y = \qquad\qquad I_z =$$
$$\iiint_E (y^2 + z^2)\rho(x,\, y,\, z)dV \quad \iiint_E (x^2 + z^2)\rho(x,\, y,\, z)dV \quad \iiint_E (x^2 + y^2)\rho(x,\, y,\, z)dV$$

6) Find a triple iterated integral for the
   volume inside the cone $z^2 = x^2 + y^2$
   bounded by $z = 0$, $z = 8$.

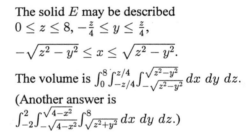

The solid E may be described
$0 \le z \le 8$, $-\frac{z}{4} \le y \le \frac{z}{4}$,

$-\sqrt{z^2 - y^2} \le x \le \sqrt{z^2 - y^2}$.

The volume is $\int_0^8 \int_{-z/4}^{z/4} \int_{-\sqrt{z^2-y^2}}^{\sqrt{z^2-y^2}} dx\, dy\, dz$.
(Another answer is
$\int_{-2}^2 \int_{-\sqrt{4-x^2}}^{\sqrt{4-x^2}} \int_{\sqrt{z^2+y^2}}^8 dx\, dy\, dz$.)

7) A solid has density xyz and is
   described as bounded by the cylinder
   $x = z^2$ and the planes $y = 0$, $z = 0$,
   $y = 5$, $z = 2$.

   a) Find an expression for the
      z-coordinate of the center mass.

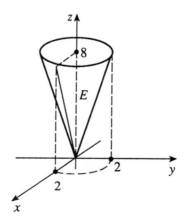

The solid E may be described as
$0 \le y \le 5, 0 \le z \le 2, 0 \le x \le z^2$.
The mass $m = \iiint\limits_{E} xyz \; dV$

$$= \int_0^5 \int_0^2 \int_0^{z^2} xyz \; dx \; dz \; dy.$$

$$M_{xy} = \iiint\limits_{E} z(xyz)dV$$

$$= \int_0^5 \int_0^2 \int_0^{z^2} xyz^2 \; dx \; dz \; dy.$$

$$\overline{z} = \frac{M_{xy}}{m}.$$

b) Find an expression for the moment of inertia about the x-axis.

$$I_x = \iiint\limits_{E} (y^2 + z^2)xyz \; dV$$

$$= \int_0^5 \int_0^2 \int_0^{z^2} (xy^3 z + xyz^3)dx \; dz \; dy.$$

Triple Integrals in Cylindrical and Spherical Coordinates

Concepts to Master

A. Triple Integrals as iterated integrals in cylindrical coordinates
B. Triple integrals as iterated integrals in spherical coordinates

---

Summary and Focus Questions

A. If f is continuous on a solid region E and E may be described in cylindrical coordinates by $\alpha \leq \theta \leq \beta, h_1(\theta) \leq r \leq h_2(\theta)$, $\phi_1(x, y) \leq z \leq \phi_2(x, y)$, then

$$\iiint_E f(x, y, z)dV = \int_\alpha^\beta \int_{h_1(\theta)}^{h_2(\theta)} \int_{\phi_1(r\cos\theta, r\sin\theta)}^{\phi_2(r\cos\theta, r\sin\theta)} f(r\cos\theta, r\sin\theta, z)r \; dz \; dr \; d\theta.$$

1) Evaluate $\iiint_E 2 \; dV$ where E is the solid between the paraboloids
$$z = 4x^2 + 4y^2$$
$$z = 80 - x^2 - y^2.$$

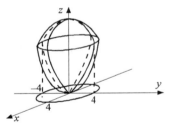

The paraboloids intersect in the circle $4x^2 + 4y^2 = 80 - x^2 - y^2$.
This simplifies to $x^2 + y^2 = 16$.
E may be described in cylindrical coordinates as $0 \leq \theta \leq 2\pi$, $0 \leq r \leq 4$, $4x^2 + 4y^2 \leq z \leq 80 - x^2 - y^2$.
Thus $4r^2 \leq z \leq 80 - r^2$.

135

$$\iiint\limits_{E} 2 \, dV = \int_0^{2\pi} \int_0^4 \int_{4r^2}^{80-r^2} 2r \, dz \, dr \, d\theta$$

$$= \int_0^{2\pi} \int_0^4 2zr \Big|_{z=4r^2}^{80-r^2} dr \, d\theta$$

$$= \int_0^{2\pi} \int_0^4 (160r - 10r^3) dr \, d\theta$$

$$= \int_0^{2\pi} \left(80r^2 - \tfrac{5}{2}r^4\right)\Big|_{r=0}^4 d\theta$$

$$= \int_0^{2\pi} 640 \, d\theta = 640\theta \Big|_{\theta=0}^{2\pi} = 1280\pi.$$

B. If f is continuous on a solid region E and E may be described in spherical coordinates by $\alpha \le \theta \le \beta,\, c \le \phi \le d,\, g_1(\theta, \phi) \le \rho \le g_2(\theta, \phi)$, then
$$\iiint\limits_{E} f(x, y, z) dV$$
$$= \int_c^d \int_\alpha^\beta \int_{g_1(\theta, \phi)}^{g_2(\theta, \phi)} f(\rho \sin\phi \cos\theta,\, \rho \sin\phi \sin\theta,\, \rho \cos\phi)\rho^2 \sin\phi \, d\rho \, d\theta \, d\phi.$$

2) Write as an iterated integral:
   $\iiint\limits_{E} (10 - x^2 - y^2 - z^2) dV$, where E
   is the top half of the sphere
   $x^2 + y^2 + z^2 = 9$.

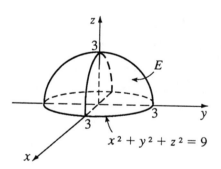

$x^2 + y^2 + z^2 = 9$

E may be described in spherical
coordinates as
$0 \le \rho \le 3,\, 0 \le \theta \le 2\pi,\, 0 \le \phi \le \frac{\pi}{2}.$
$\iiint\limits_{E} (10 - x^2 - y^2 - z^2) dV$
$= \int_0^{\pi/2} \int_0^{2\pi} \int_0^3 (10 - \rho^2)\rho^2 \sin\phi \, d\rho \, d\theta \, d\phi.$

3) Which system, spherical or cylindrical, would be more appropriate to evaluate $\iiint\limits_E x \, dV$ where E is the solid between the cone $z = \sqrt{x^2 + y^2}$ and the plane $z = 4$?

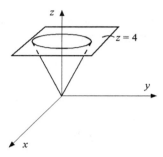

Cylindrical, since E may be written as
$0 \le \theta \le 2\pi$,
$0 \le r \le 2$,
$\sqrt{x^2 + y^2} \le z \le 4$.

Change of Variables in Multiple Integrals

Concepts to Master

A. C^1 transformations; Jacobians in 2 and 3 dimensions
B. Change of variable for double and triple integrals

Summary and Focus Questions

A. A <u>transformation</u> in the plane is a function T from the uv-plane to the xy-plane
$$T(u,\ v) = (x,\ y), \text{ where } x = g(u,\ v) \text{ and } y = h(u,\ v).$$

T is $\underline{C^1}$ means g and h have continuous first derivatives.
Similarly, a transformation in space is a function T from the uvw-space to xyz-space.
For a C^1 transformation T from uv to xy, the <u>Jacobian of T</u> is the determinant

$$\frac{\partial(x, y)}{\partial(u, v)} = \begin{vmatrix} \frac{\partial x}{\partial u} & \frac{\partial x}{\partial v} \\ \frac{\partial y}{\partial u} & \frac{\partial y}{\partial v} \end{vmatrix}.$$

In three dimensions the Jacobian is

$$\frac{\partial(x, y, z)}{\partial(u, v, w)} = \begin{vmatrix} \frac{\partial x}{\partial u} & \frac{\partial x}{\partial v} & \frac{\partial x}{\partial w} \\ \frac{\partial y}{\partial u} & \frac{\partial y}{\partial v} & \frac{\partial y}{\partial w} \\ \frac{\partial z}{\partial u} & \frac{\partial z}{\partial v} & \frac{\partial z}{\partial w} \end{vmatrix}.$$

1) Find the Jacobian of each transformation:

a) $x = 2u + v$
$y = u - 2v$

$$\frac{\partial(x, y)}{\partial(u, v)} = \begin{vmatrix} 2 & 1 \\ 1 & -2 \end{vmatrix} = (2)(-2) - (1)(1)$$
$$= -5.$$

b) $x = uv$
$y = u + v$

$$\frac{\partial(x, y)}{\partial(u, v)} = \begin{vmatrix} v & u \\ 1 & 1 \end{vmatrix} = v - u.$$

c) $x = 2u + v$
     $y = uw$
     $z = u^2 + v^2 + w^2$

$$\frac{\partial(x, y, z)}{\partial(u, v, w)} = \begin{vmatrix} 2 & 1 & 0 \\ w & 0 & u \\ 2u & 2v & 2w \end{vmatrix}$$

$$= 2\begin{vmatrix} 0 & u \\ 2v & 2w \end{vmatrix} - 1\begin{vmatrix} w & u \\ 2u & 2w \end{vmatrix}$$

$$+ 0\begin{vmatrix} w & 0 \\ 2u & 2v \end{vmatrix}$$

$$= 2(-2uv) - 1(2w^2 - 2u^2) + 0$$
$$= 2u^2 - 4uv - 2w^2.$$

2) Find the region S that maps to the
    region R under the transformation T:
    $$x = \tfrac{2}{3}(v - u)$$

    $$y = \tfrac{1}{3}(u + 2v)$$
    R is the region following:

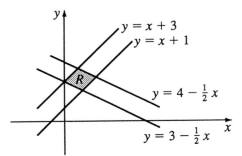

We find T^{-1} by solving for u and v.
Multiplying each of
$x = \tfrac{2}{3}(v - u)$ and $y = \tfrac{1}{3}(u + 2v)$ by 3
gives $3x = 2v - 2u$
         $3y = 2v + u.$
Subtracting the equations gives
$3x - 3y = -3u.$ Thus $u = y - x.$
Then $3x = 2v - 2(y - x),$ so $v = \tfrac{1}{2}x + y.$
The four lines are transformed by T^{-1} from
the xy-plane to the uv-plane as follows:

xy	uv
$y = 4 - \tfrac{1}{2}x$	$v = 4$
$y = 3 - \tfrac{1}{2}x$	$v = 3$
$y = x + 3$	$u = 3$
$y = x + 1$	$u = 1$

Thus $S = \{(u, v) | 1 \le u \le 3, 3 \le v \le 4\}$
$= [1, 3] \times [3, 4]$ and S is mapped to R by
T.

B. Let T be a one-to-one C^1 transformation that maps Type I or II region S in the uv-plane onto Type I or II region R in the xy-plane. If T has a nonzero Jacobian and f is continuous, then

$$\iint\limits_R f(x, y)dx\ dy = \iint\limits_S f(x(u, v), y(u, v))\left|\frac{\partial(x, y)}{\partial(u, v)}\right|du\ dv.$$

A similar result holds for transformations in three dimensions.

3) Use question 2 to change the variable in $\iint\limits_R (6x + 3y)dA$ and evaluate.

With the transformation T:

$x = \frac{2}{3}(v - u),\ y = \frac{1}{3}(u + 2v)$

$T(S) = R$ where $S = [1, 3] \times [3, 4]$.

The Jacobian of T is

$$\begin{vmatrix} \frac{\partial x}{\partial u} & \frac{\partial x}{\partial v} \\ \frac{\partial y}{\partial u} & \frac{\partial y}{\partial v} \end{vmatrix} = \begin{vmatrix} -\frac{2}{3} & \frac{2}{3} \\ \frac{1}{3} & \frac{2}{3} \end{vmatrix} = -\frac{4}{9} - \frac{2}{9} = -\frac{2}{3}$$

Thus $\iint\limits_R (6x + 3y)dA$

$$= \iint\limits_S \left(6\left[\frac{2}{3}(v - u)\right]\right.$$

$$\left. + 3\left[\frac{1}{3}(u + 2v)\right]\right)\left|-\frac{2}{3}\right|du\ dv$$

$$= \iint\limits_S (4v - 2u)du\ dv$$

$$= \int_3^4 \int_1^3 (4v - 2u)du\ dv$$

$$= \int_3^4 (4uv - u^2)\Big|_{u=1}^3 dv$$

$$= \int_3^4 (8v - 8)dv$$

$$= (4v^2 - 8v)\Big|_{v=3}^4 = 20.$$

4) Use a change of variable to evaluate $\iint\limits_R (x^2 - y^2)dx\ dy$ where R is the shaded region.

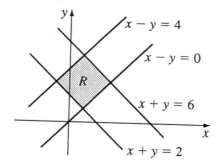

The region is bounded by lines whose equations include the expressions $x + y$ and $x - y$.

We let $u = x + y,\ v = x - y$.

Then $2 \leq u \leq 6,\ 0 \leq v \leq 4$.

We must write x, y in terms of u, v :

$u = x + y$

$v = x - y$

$u + v = 2x$, so $x = \frac{u+v}{2}$.

Thus $u = \frac{u+v}{2} + y$, so $y = \frac{u-v}{2}$.

The integrand

$f(x,\ y) = x^2 - y^2$

$\qquad = \left(\frac{u+v}{2}\right)^2 - \left(\frac{u-v}{2}\right)^2 = uv$.

Finally, the Jacobian is

$$\begin{vmatrix} \frac{\partial x}{\partial u} & \frac{\partial x}{\partial v} \\ \frac{\partial y}{\partial u} & \frac{\partial y}{\partial v} \end{vmatrix} = \begin{vmatrix} \frac{1}{2} & \frac{1}{2} \\ \frac{1}{2} & -\frac{1}{2} \end{vmatrix} = -\frac{1}{4} - \frac{1}{4} = -\frac{1}{2}.$$

Thus

$\iint\limits_{R} (x^2 - y^2)\,dx\,dy$

$\qquad = \int_2^6 \int_0^4 uv \left|-\frac{1}{2}\right| dv\,du$

$\qquad = \int_2^6 \int_0^4 \frac{uv}{2}\,dv\,du$

$\qquad = \int_2^6 \frac{uv^2}{4}\Big|_{v=0}^{4}\,du$

$\qquad = \int_2^6 4u\,du = 2u^2\Big|_{u=2}^{6} = 64.$

14

Vector Calculus

"BUT WE JUST DON'T HAVE THE TECHNOLOGY TO CARRY IT OUT."

Cartoons courtesy of Sidney Harris. Used by permission.

Vector Fields

Concepts to Master

Vector fields in two and three dimensions; Conservative vector fields

Summary and Focus Questions

A <u>vector field on R^2</u> is a function **F** that assigns to each point in a domain $D \subseteq R^2$ a two-dimensional vector $\mathbf{F}(x, y)$. When points in D are thought of as vectors, $\mathbf{x} = (x, y)$, we write $\mathbf{F}(\mathbf{x})$. Also, the vector $\mathbf{F}(x, y)$ may be written in terms of its <u>scalar field</u> component functions:

$$\mathbf{F}(x, y) = P(x, y)\mathbf{i} + Q(x, y)\mathbf{j}$$

A vector field can be represented by drawing the plane, choosing some representative points and sketching the associated vectors at each of those points. For example, let $\mathbf{F}(x, y) = x\mathbf{i} + 2\mathbf{j}$. The component functions are $P(x, y) = x$ and $Q(x, y) = 2$. Here is a sketch of F with four domain points $(0, 0)$, $(1, -4)$, $(3, 1)$, and $(-1, -1)$ and their associated vectors:

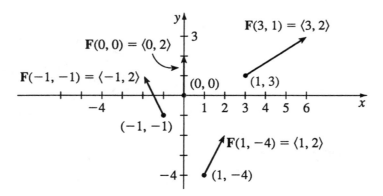

The concepts are similar in three dimensions. For example,
$\mathbf{F}(x, y, z) = (y + z)\mathbf{i} + (x + z)\mathbf{j} + (x + y)\mathbf{k}$ is a three dimensional vector field with component functions $P(x, y, z) = y + z$, $Q(x, y, z) = x + z$, $R(x, y, z) = x + y$.

For a scalar function f, the gradient, ∇f, is a vector field. In two dimensions
$$\nabla f(x, y) = f_x(x, y)\mathbf{i} + f_y(x, y)\mathbf{j}.$$

A vector field $\mathbf{F}$ is <u>conservative</u> means $\mathbf{F} = \nabla f$ for some scalar function f called
the <u>potential field of $\mathbf{F}$</u>. For example, $\mathbf{F}(x, y) = 2xy\mathbf{i} + x^2\mathbf{j}$ is conservative
because $\mathbf{F} = \nabla f$, where $f(x, y) = x^2 y$ (Note: $f_x = 2xy$ and $f_y = x^2$, the
components of $\mathbf{F}$). We will have a means for checking for conservativeness in
Section 14.3.

1) For the vector field
    $\mathbf{F}(x, y) = |x|\mathbf{i} + |y|\mathbf{j}$, sketch the
    images of $(0, 0)$, $(-2, 0)$, $(-3, 1)$,
    $(1, 2)$, $(2, -2)$.

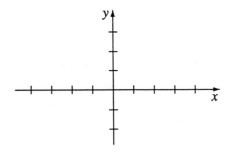

2) Find the gradient vector field of

    a) $f(x, y) = x^2 + 3xy$

    $\nabla f = (2x + 3y)\mathbf{i} + 3x\mathbf{j}.$

    b) $f(x, y, z) = x^2 y + xz^2 + 2yz$

    $\nabla f = (2xy + z^2)\mathbf{i} + (x^2 + 2z)\mathbf{j}$
    $\qquad + (2xz + 2y)\mathbf{k}$

3) If $\mathbf{F}$ is a conservative vector field, then
    _____ for some scalar function f.

    $\nabla f = \mathbf{F}.$

Line Integrals

Concepts to Master

A. Line integral in two and three dimensions; Piecewise smooth curve; Orientation of curves

B. Line integrals of vector fields; Work

Summary and Focus Questions

A. Let C be a smooth curve given by $\mathbf{r}(t) = x(t)\mathbf{i} + y(t)\mathbf{j}$, $t \in [a, b]$, and let f be a function of two variables. The <u>line integral of f along C</u> is:

$$\int_C f(x, y)ds = \lim_{\|P\| \to 0} \sum_{i=1}^{n} f(x_i^*, y_i^*)\Delta s_i$$

where $P = \{t_0, t_1, t_2, \ldots, t_n\}$, is a partition of $[a, b]$, (which in turn divides C into subarcs), Δs_i is the length of the i-th subarc, and each (x_i^*, y_i^*) is a point from the i-th subarc (which corresponds to t_i^* in the i-th subinterval of P.

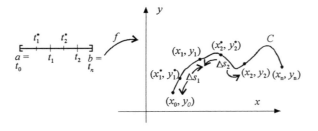

If f is continuous,

$$\int_C f(x, y)ds = \int_a^b f(x(t), y(t))\sqrt{\left(\frac{dx}{dt}\right)^2 + \left(\frac{dy}{dt}\right)^2}\, dt.$$

Line integrals in 3-dimensional space are defined and computed similarly. For example,

$$\int_C f(x, y, z)ds = \int_a^b f(x(t), y(t), z(t))\sqrt{\left(\frac{dx}{dt}\right)^2 + \left(\frac{dy}{dt}\right)^2 + \left(\frac{dz}{dt}\right)^2}\, dt.$$

A curve C is <u>piecewise smooth</u> if C
is a union of a finite number of
smooth subarcs.

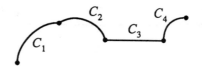

If C is piecewise smooth and composed of smooth curves C_1 and C_2 then,

$$\int_C f \ ds = \int_{C_1} f \ ds + \int_{C_2} f \ ds.$$

The line integral of f along C <u>with respect to x</u> is

$$\int_C f(x, \ y)dx = \int_a^b f(x(t), \ y(t))x'(t)dt$$

and <u>with respect to y</u> is

$$\int_C f(x, \ y)dy = \int_a^b f(x(t), \ y(t))y'(t)dt.$$

A parametrization $x = x(t), \ y = y(t), \ t \in [a, \ b]$ of a curve C determines an
<u>orientation</u> or direction along the curve as t increases from a to b. The curve
$-C$ is the same set of points as C but with the opposite orientation.

1)   Evaluate $\int_C y \ ds$, where C is given by
    $x(t) = t, \ y(t) = t^3, \ 0 \leq t \leq \sqrt[4]{7}$.

$x'(t) = 1$ and $y'(t) = 3t^2$, so

$$\int_C y \ ds = \int_0^{\sqrt[4]{7}} t^3 \sqrt{1^2 + (3t^2)^2} \ dt$$

$$= \int_0^{\sqrt[4]{7}} t^3 \sqrt{1 + 9t^4} \ dt$$

$$= \tfrac{1}{54}(1 + 9t^4)^{3/2} \Big|_{t=0}^{\sqrt[4]{7}} = \tfrac{511}{54}.$$

2)   Set up definite integrals for
    $\int_C (x - y)dx$ and $\int_C (x - y)dy$ where
    C is the curve given by $x = 1 + e^t$,
    $y = 1 - e^{-t}, \ 0 \leq t \leq 1$.

$x'(t) = e^t, \ y'(t) = e^{-t}$

$$\int_C (x - y)dx = \int_0^1 (1 + e^t - (1 - e^{-t}))e^t \ dt$$

$$= \int_0^1 (e^{2t} + 1)dt.$$

$$\int_C (x - y)dy = \int_0^1 (1 + e^t - (1 - e^{-t}))e^{-t} \ dt$$

$$= \int_0^1 (1 + e^{-2t})dt.$$

3) Write a definite integral for $\int_C xyz\ ds$ where C is the helix $x = t$, $y = \sin t$, $z = \cos t$, $0 \le t \le \pi$.

$$\int_0^\pi (t \sin t \cos t)\sqrt{1^2 + \cos^2 t + \sin^2 t}\ dt$$
$$= \int_0^\pi \sqrt{2}\, t \sin t \cos t\ dt = \int_0^\pi \frac{t \sin 2t}{\sqrt{2}}\ dt.$$

B. If **F** is a continuous vector field on a smooth curve C given by $\mathbf{r}(t)$, $t \in [a,\ b]$, the <u>line integral of **F** along C</u> is

$$\int_C \mathbf{F}\ d\mathbf{r} = \int_a^b \mathbf{F}(\mathbf{r}(t)) \cdot \mathbf{r}'(t)dt.$$

This may also be written as $\int_C \mathbf{F} \cdot \mathbf{T}\ ds$ where **T** is the unit tangent vector function for C, and $\mathbf{F} \cdot \mathbf{T}$ is the dot product.

A line integral of a vector field $\mathbf{F} = P\mathbf{i} + Q\mathbf{j} + R\mathbf{k}$ may be written in terms of line integrals of its components:

$$\int_C \mathbf{F} \cdot d\mathbf{r} = \int_C P\ dx + Q\ dy + R\ dz.$$

$\left(\int_C P\ dx + Q\ dy + R\ dz \text{ is shorthand for } \int_C P\ dx + \int_C Q\ dy + \int_C R\ dz. \right)$

If **F** is interpreted as a force field, then $\int_C \mathbf{F} \cdot d\mathbf{r}$ is the work done by the force **F** in moving a particle along the curve C.

4) Evaluate $\int_C \mathbf{F} \cdot d\mathbf{r}$ where C is the line segment from $(0,\ 0,\ 0)$ to $(1,\ 3,\ 2)$ and
$\mathbf{F}(x,\ y,\ z) = (y + z)\mathbf{i}$
$\qquad + (2x + z)\mathbf{j} + (x + 3y)\mathbf{k}.$

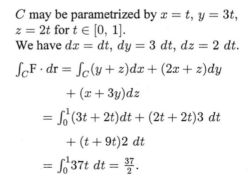

C may be parametrized by $x = t$, $y = 3t$, $z = 2t$ for $t \in [0,\ 1]$.
We have $dx = dt$, $dy = 3\ dt$, $dz = 2\ dt$.

$\int_C \mathbf{F} \cdot d\mathbf{r} = \int_C (y + z)dx + (2x + z)dy$
$\qquad\qquad + (x + 3y)dz$
$= \int_0^1 (3t + 2t)dt + (2t + 2t)3\ dt$
$\qquad + (t + 9t)2\ dt$
$= \int_0^1 37t\ dt = \frac{37}{2}.$

5) In the force field
$\mathbf{F}(x, y) = x\mathbf{i} + (x + y)\mathbf{j}$, find the
amount of work done moving a
particle along the parabola $y = x^2$
from $(1, 1)$ to $(2, 4)$.

The parabolic curve C may be described by
$x = t$, $y = t^2$, $t \in [1, 2]$.
$x'(t) = 1$, $y'(t) = 2t$.
$dx = dt$, $dy = 2t\ dt$.
The work done is
$\int_C \mathbf{F} \cdot d\mathbf{r} = \int_C x\ dx + (x + y)dy$

$= \int_1^2 t(1)dt + (t + t^2)2t\ dt$

$= \int_1^2 (t + 2t^2 + 2t^3)dt = \frac{41}{3}.$

The Fundamental Theorem for Line Integrals

Concepts to Master

A. Line integrals of gradients
B. Closed path; Simple path; Connected; Simply-connected; Independence of path; Test for path independence

---

Summary and Focus Questions

A. If C is a piecewise smooth curve given by $\mathbf{r}(t)$, $a \leq t \leq b$ and f is a differentiable function with ∇f continuous on C, then

$$\int_C \nabla f \cdot d\mathbf{r} = f(\mathbf{r}(b)) - f(\mathbf{r}(a)).$$

1) Let $f(x, y) = x^2 + xy + 2y^2$ and C be any smooth curve from $(0, 1)$ to $(3, 0)$. Find $\int_C \nabla f \cdot d\mathbf{r}$.

$$\int_C \nabla f \cdot d\mathbf{r} = f(3, 0) - f(0, 1)$$
$$= 9 - 2 = 7.$$

2) If C is a smooth curve given by $\mathbf{r}(t)$ with $\mathbf{r}(a) = \mathbf{r}(b)$ and f is differentiable and ∇f is continuous on C, then $\int_C \nabla f \cdot d\mathbf{r} = \underline{\quad\quad}$.

0.

B. A curve C given by r(t), $a \leq t \leq b$ is <u>closed</u> means r$(a) =$ r(b).

A curve C is <u>simple</u> means C does not intersect itself except, perhaps, at its endpoints.

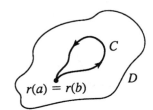

149

A domain D is <u>open</u> if every point is an interior point. D is <u>connected</u> if any two points of D can be connected by a path entirely in D. D is <u>simply-connected</u> means every simple closed curve in D encloses only points of D.

For a continuous vector function $\mathbf{F}$ with domain D, $\int_C \nabla f \cdot d\mathbf{r}$ is <u>independent of path</u> means its value depends only upon the endpoints A and B and not on any particular path from A to B.

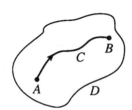

The results in part **A** of this section say that $\int_C \nabla f \cdot d\mathbf{r}$ is independent of path. Conversely, under certain conditions, every independent of path line integral is a gradient line integral. Let $\mathbf{F}$ be continuous on an open connected region D. If $\int_C \nabla f \cdot d\mathbf{r}$ is independent of path, then $\mathbf{F}$ is conservative (i.e., $\mathbf{F} = \nabla f$, for some differentiable f).

For an open simply-connected region D, if $\frac{\partial P}{\partial y}$ and $\frac{\partial Q}{\partial x}$ are continuous on D, then $\mathbf{F} = P\mathbf{i} + Q\mathbf{j}$ is the gradient of some function f if and only if $\frac{\partial P}{\partial y} = \frac{\partial Q}{\partial x}$. Thus $\frac{\partial P}{\partial y} = \frac{\partial Q}{\partial x}$ is a test of whether $\mathbf{F}$ is conservative (and thus path independent). If $\mathbf{F} = P\mathbf{i} + Q\mathbf{j} = \nabla f$, to reconstruct f, integrate $\int P \, dx$, differentiate the result with respect to y, and then compare with Q. Or you can compare P with $\frac{\partial}{\partial x}(\int Q \, dy)$.

3) True, False:
   The curve below is simple.

False.

4) True, False:
The region below is simply connected.

5) If **F** is conservative and C is a smooth closed curve then $\int_C \mathbf{F} \cdot d\mathbf{r} = \underline{\hspace{1cm}}$.

6) Suppose C_1 and C_2 are two distinct paths in a region D from point A to point B. If $\int_{C_1} \mathbf{F} d\mathbf{r} = \int_{C_2} \mathbf{F} d\mathbf{r}$, can we conclude the line integral is independent of path?

7) Is $\mathbf{F}(x, y) = (2x + y)\mathbf{i} + (x + 8y)\mathbf{j}$ the gradient of some function?

8) Find the function f whose gradient is $\nabla f = (y^2 + 1)\mathbf{i} + (2xy + 3y^2)\mathbf{j}$.

True.

0. (For some f, $\mathbf{F} = \nabla f$, so $\int_C \mathbf{F} \cdot d\mathbf{r} = \int_C \nabla f \cdot d\mathbf{r} = 0$ since the curve is closed.)

No, independence requires every line integral along paths in D from A to B to have the same value.

Yes. For $P(x, y) = 2x + y$ and $Q(x, y) = x + 8y$, $\frac{\partial P}{\partial y} = 1 = \frac{\partial Q}{\partial x}$, so $\mathbf{F} = \nabla f$ for some f.

Let $P(x, y) = y^2 + 1$ and $Q(x, y) = 2xy + 3y^2$.
$f(x, y) = \int P\ dx$
$\qquad = \int (y^2 + 1)dx = xy^2 + x + h(y)$,
where h is a function of y.
$\frac{\partial f}{\partial y} = 2xy + h'(y)$. Comparing to Q,
$h'(y) = 3y^2$ so $h(y) = y^3 + K$.
Thus $f(x, y) = xy^2 + x + y^3 + K$.

9) Let C be the parabola $x = y^2$ from
(1, 1) to (4, 2). Evaluate
$\int_C (2x + 3y^2)dx + (6xy + 10)dy$.

Let $P(x, y) = 2x + 3y^2$ and
$Q(x, y) = 6xy + 10$.

Since $\frac{\partial P}{\partial y} = 6y = \frac{\partial Q}{\partial x}$, $P\mathbf{i} + Q\mathbf{j}$ is a
gradient of some function $f(x, y)$.
$f(x, y) = \int (2x + 3y^2)dx$
$\qquad = x^2 + 3xy^2 + h(y)$.
$\frac{\partial f}{\partial y} = 6xy + h'(y)$.
Comparing with $Q(x, y)$, $h'(y) = 10$,
so $h(y) = 10y + c$. We choose $c = 0$.
Thus $f(x, y) = x^2 + 3xy^2 + 10y$ and
$\int_C P\, dx + Q\, dy = f(4, 2) - f(1, 1)$
$\qquad = 84 - 14 = 70$.
(Note that it did not matter that C was a
parabola.)

10) Evaluate $\int_C dx + z\, dy + y\, dz$ where C
is any path from (0, 1, 1) to
(3, 1, 6).

Let $P(x, y, z) = 1$, $Q(x, y, z) = z$,
$R(x, y, z) = y$
$f(x, y, z) = \int P\, dx = x + C(y, z)$.
$f_y = C_y(y, z) = Q = z$.
So $C(y, z) = yz + D(z)$.
Thus $f(x, y, z) = x + yz + D(z)$.
$f_z = y + D'(z) = R = y$. Thus
$D'(z) = 0$, so $D(z) = K$. We use $K = 0$.
Therefore $f(x, y, z) = x + yz$ and
$\int_C dx + z\, dy + y\, dz = (x + yz)\Big|_{(0,1,1)}^{(3,1,6)}$
$\qquad = (3 + 6) - (0 + 1) = 8$.

Green's Theorem

Concepts to Master

A. Positive orientation of a plane curve; Green's Theorem
B. Line integrals for areas

Summary and Focus Questions

A. A simple closed curve C, given by $\mathbf{r}(t)$, $a \le t \le b$, has <u>positive orientation</u> if the traversal from $\mathbf{r}(a)$ to $\mathbf{r}(b)$ along C is counterclockwise.

The line integral about a curve C with positive orientation is denoted $\oint_C \mathbf{F} \cdot d\mathbf{r}$.

<u>Green's Theorem</u>: Let C be a piecewise smooth, simple closed curve with positive orientation about a region D. If $P(x, y)$ and $Q(x, y)$ have continuous partials in an open region containing D, then
$$\oint_C P\, dx + Q\, dy$$
$$= \iint_D \left(\frac{\partial Q}{\partial x} - \frac{\partial P}{\partial y} \right) dA.$$

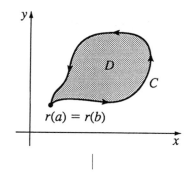

Green's Theorem is an important connection between line integrals and double integrals.

1) Place an arrow at the point x to indicate a positive orientation to the curve C.

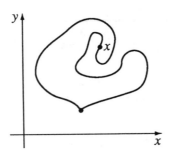

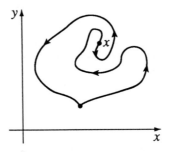

2) Evaluate using Green's Theorem:

a) $\oint_C xy \, dx + y^2 \, dy$, where C is the triangle formed by $(0, 0)$, $(1, 1)$, and $(0, 1)$.

Let $P(x, y) = xy$, $Q(x, y) = y^2$.
Then $\frac{\partial P}{\partial y} = x$ and $\frac{\partial Q}{\partial x} = 0$.
C is pictured below and encloses the region D: $0 \le x \le 1$, $x \le y \le 1$.

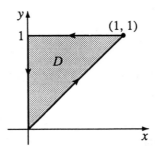

The line integral becomes
$$\iint_D (0 - x) dx \, dy = \int_0^1 \int_x^1 -x \, dy \, dx$$
$$= \int_0^1 (x^2 - x) dx = -\tfrac{1}{6}.$$

b) $\oint_C (6xy^2)dx + (6x^2y)dy$, where
C is the triangle in part a).

For $P(x,\,y) = 6xy^2$ and $Q(x,\,y) = 6x^2y$,
$\frac{\partial P}{\partial y} = 12xy = \frac{\partial Q}{\partial x}$.
Thus $\frac{\partial Q}{\partial x} - \frac{\partial P}{\partial y} = 0$,
so $\oint_C P\ dx + Q\ dy = \iint_D 0\ dA = 0$.

B. By selecting P and Q such that
$\frac{\partial Q}{\partial x} - \frac{\partial P}{\partial y} = 1$ (there are many
choices), line integrals may be used
to find areas.
Area of $D = \oint_C x\ dy$
$= -\oint_C y\ dx$
$= \frac{1}{2}\oint x\ dy - y\ dx$.

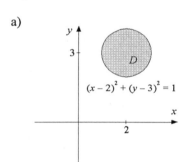

3) Find a line integral expression for the
region D:

a)

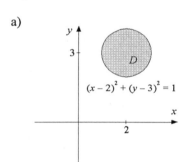

We know the answer will be π since D is a
circle of radius 1. The boundary C may be
described as $x = 2 + \cos\theta$, $y = 3 + \sin\theta$,
$0 \le \theta \le 2\pi$. The area of D is
$\frac{1}{2}\int x\ dy - y\ dx$
$= \frac{1}{2}\int_0^{2\pi}(2 + \cos\theta)\cos\theta$
$\qquad - (3 + \sin\theta)(-\sin\theta)d\theta$
$= \frac{1}{2}\int_0^{2\pi}(2\cos\theta - 3\sin\theta + 1)d\theta$
$= \frac{1}{2}(2\sin\theta + 3\cos\theta + \theta)\Big|_0^{2\pi} = \pi$.

Curl and Divergence

Concepts to Master

A. Del operator; Curl of a vector field
B. Divergence of a vector field
C. Vector forms of Green's Theorem

Summary and Focus Questions

A. The operations in this section are rather like a form of differentiation where the result is either a vector field (curl **F**) or a scalar field (div **F**).

The $\underline{\nabla \text{ (del)}}$ operator is $\nabla = \mathbf{i}\frac{\partial}{\partial x} + \mathbf{j}\frac{\partial}{\partial y} + \mathbf{k}\frac{\partial}{\partial z}$.

The $\underline{\text{curl of } \mathbf{F}} = P\mathbf{i} + Q\mathbf{j} + R\mathbf{k}$ is

$$\text{curl } \mathbf{F} = \nabla \times \mathbf{F} = \begin{vmatrix} \mathbf{i} & \mathbf{j} & \mathbf{k} \\ \frac{\partial}{\partial x} & \frac{\partial}{\partial y} & \frac{\partial}{\partial z} \\ P & Q & R \end{vmatrix}$$

$$= \left(\frac{\partial R}{\partial y} - \frac{\partial Q}{\partial z}\right)\mathbf{i} + \left(\frac{\partial P}{\partial z} - \frac{\partial R}{\partial x}\right)\mathbf{j} + \left(\frac{\partial Q}{\partial x} - \frac{\partial P}{\partial y}\right)\mathbf{k}.$$

The curl **F** measures a rate of change of circulation -- if **F** is the velocity of a fluid flow, then curl $\mathbf{F}(x, y, z)$ measures the circulation per unit area orthogonal to $\mathbf{F}(x, y, z)$ at (x, y, z). If curl $\mathbf{F} = 0$ (irrotation) at a point, there is no circulation about the point as fluid flows through the point with force **F**.

If $f(x, y, z)$ has continuous second partials, $\text{curl}(\nabla f) = \mathbf{0}$, the zero vector. Conversely, if $\mathbf{F}(x, y, z)$ has components with continuous partial derivatives and curl $\mathbf{F} = \mathbf{0}$ then **F** is conservative.

1) Find the curl of
$$\mathbf{F}(x,\ y,\ z) = xy^2\mathbf{i} + x^2z\mathbf{j} + yz^2\mathbf{k}.$$

Let $P(x,\ y,\ z) = xy^2$, $Q(x,\ y,\ z) = x^2z$,
and $R(x,\ y,\ z) = yz^2$.
$\frac{\partial P}{\partial y} = 2xy$, $\frac{\partial P}{\partial z} = 0$, $\frac{\partial Q}{\partial x} = 2xz$,

$\frac{\partial Q}{\partial z} = x^2$, $\frac{\partial R}{\partial x} = 0$, $\frac{\partial R}{\partial y} = z^2$.
curl $\mathbf{F} = (z^2 - x^2)\mathbf{i} + (0 - 0)\mathbf{j}$
$\qquad\quad + (2xz - 2xy)\mathbf{k}$
$\qquad = (z^2 - x^2)\mathbf{i} - (2xz - 2xy)\mathbf{k}.$

2) Use the curl $\mathbf{F}$ to determine whether
$$\mathbf{F}(x,\ y,\ z) = xy\mathbf{i} + z^2\mathbf{j} + xz\mathbf{k}$$
is conservative.

$\frac{\partial P}{\partial y} = x$, $\frac{\partial P}{\partial z} = 0$, $\frac{\partial Q}{\partial x} = 0$,

$\frac{\partial Q}{\partial z} = 2z$, $\frac{\partial R}{\partial x} = z$, $\frac{\partial R}{\partial y} = 0$.
curl $\mathbf{F} = (0 - 2z)\mathbf{i} + (0 - z)\mathbf{j} + (0 - x)\mathbf{k}$
$\qquad = -2z\mathbf{i} - z\mathbf{j} - x\mathbf{k}$, not the zero
vector.
$\mathbf{F}$ is not conservative.

3) Show that $\mathbf{F}(x,\ y,\ z) =$
$(2x + y)\mathbf{i} + (x + 2yz^2)\mathbf{j} + 2y^2z\mathbf{k}$
is conservative and find any f such
that $\nabla f = \mathbf{F}$.

$\frac{\partial P}{\partial y} = 1$, $\frac{\partial P}{\partial z} = 0$, $\frac{\partial Q}{\partial x} = 1$,

$\frac{\partial Q}{\partial z} = 4yz$, $\frac{\partial R}{\partial x} = 0$, $\frac{\partial R}{\partial y} = 4yz$.
curl $\mathbf{F} = (4yz - 4yz)\mathbf{i} + (0 - 0)\mathbf{j}$
$\qquad\quad + (1 - 1)\mathbf{k} = \mathbf{0}$
so $\mathbf{F}$ is conservative.
$f_x = 2x + y$, $f_y = x + 2yz^2$, $f_z = 2y^2z$.
From $f_x = 2x + y$,
$f(x,\ y,\ z) = x^2 + xy + C(y,\ z).$
$f_y = x + C_y(y,\ z) = x + 2yz^2.$
Thus $C_y(y,\ z) = 2yz^2$,
so $C(y,\ z) = y^2z^2 + K(z).$
Hence $f(x,\ y,\ z) = x^2 + xy + y^2z^2 + K(z).$
Now $f_z = 2y^2z + K_z(z) = 2y^2z$, so
$K_z(z) = 0.$ Choose $K(z) = 0.$
Therefore $f(x,\ y,\ z) = x^2 + xy + y^2z^2.$

B. For $\mathbf{F} = P\mathbf{i} + Q\mathbf{j} + R\mathbf{k}$, <u>the divergence of $\mathbf{F}$</u> is

$$\text{div } \mathbf{F} = \nabla \cdot \mathbf{F} = \frac{\partial P}{\partial x} + \frac{\partial Q}{\partial y} + \frac{\partial R}{\partial z}.$$

If $\mathbf{F}$ is the velocity of a fluid flow, then div $\mathbf{F}(x, y, z)$ may be interpreted as the (instantaneous) rate of change of the mass of the fluid per unit volume. Thus if div $\mathbf{F}(x, y, z) > 0$ there is a net flow out of the point (x, y, z). (The fluid is diverging.)

For $\mathbf{F} = P\mathbf{i} + Q\mathbf{j} + R\mathbf{k}$ with continuous second partial derivatives, div curl $\mathbf{F} = 0$.

4) Find div $\mathbf{F}$ for
    $$\mathbf{F}(x, y, z) = (x^2 + y)\mathbf{i}$$
    $$+ (y^2 + z)\mathbf{j} + xyz\mathbf{k}.$$

$\frac{\partial P}{\partial x} = 2x, \frac{\partial Q}{\partial y} = 2y, \frac{\partial R}{\partial z} = xy.$

div $\mathbf{F} = 2x + 2y + xy.$

5) True, False:
    curl div $\mathbf{F} = 0$.

False. For a vector field $\mathbf{F}$, div $\mathbf{F}$ is a scalar. Curl can only be computed for a vector field, so curl div $\mathbf{F}$ makes no sense.

6) For a fluid with velocity
    $\mathbf{F} = 2\mathbf{i} + y^2\mathbf{j} + (x + z)\mathbf{k}$ is the fluid
    diverging at $(1, 2, 1)$?

Yes. div $\mathbf{F} = \frac{\partial P}{\partial x} + \frac{\partial Q}{\partial y} + \frac{\partial R}{\partial z}$
$= 0 + 2y + 1 = 2y + 1.$
At $(1, 2, 1)$, div $\mathbf{F} = 2(2) + 1 = 5.$
Since div $\mathbf{F} > 0$ the fluid is diverging.

C. Using the concepts of div and curl, Green's Theorem may be expressed in vector forms:

$$\oint_C \mathbf{F} \cdot d\mathbf{r} = \iint_D (\text{curl } \mathbf{F}) \cdot \mathbf{k} \, dA$$

and

$$\oint_C \mathbf{F} \cdot \mathbf{n} \, ds = \iint \text{div } \mathbf{F}(x, y) dA$$

where $\mathbf{n}$ is the outer normal to the tangent vector $\mathbf{T}$ to C,

$$\mathbf{n}(t) = \frac{y'(t)}{|\mathbf{r}'(t)|}\mathbf{i} - \frac{x'(t)}{|\mathbf{r}'(t)|}\mathbf{j}.$$

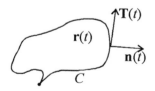

7) Write a double integral for $\oint_C \mathbf{F} \cdot d\mathbf{r}$ where C is the curve $\frac{x^2}{4} + \frac{y^2}{9} = 1$ and $\mathbf{F}(x, y) = e^{xy}\mathbf{i} - e^{xy}\mathbf{j}$.

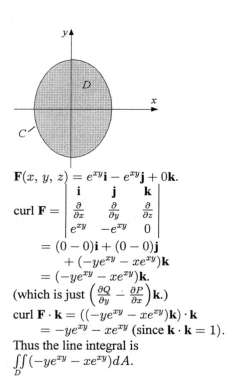

$\mathbf{F}(x, y, z) = e^{xy}\mathbf{i} - e^{xy}\mathbf{j} + 0\mathbf{k}.$

$$\text{curl } \mathbf{F} = \begin{vmatrix} \mathbf{i} & \mathbf{j} & \mathbf{k} \\ \frac{\partial}{\partial x} & \frac{\partial}{\partial y} & \frac{\partial}{\partial z} \\ e^{xy} & -e^{xy} & 0 \end{vmatrix}$$

$= (0 - 0)\mathbf{i} + (0 - 0)\mathbf{j}$
$\quad + (-ye^{xy} - xe^{xy})\mathbf{k}$
$= (-ye^{xy} - xe^{xy})\mathbf{k}.$

(which is just $\left(\frac{\partial Q}{\partial y} - \frac{\partial P}{\partial x} \right)\mathbf{k}.$)

$\text{curl } \mathbf{F} \cdot \mathbf{k} = ((-ye^{xy} - xe^{xy})\mathbf{k}) \cdot \mathbf{k}$
$\quad = -ye^{xy} - xe^{xy} \text{ (since } \mathbf{k} \cdot \mathbf{k} = 1).$

Thus the line integral is
$\iint\limits_{D} (-ye^{xy} - xe^{xy})dA.$

Parametric Surfaces and Their Areas

Concepts to Master

A. Parametrization of a surface
B. Tangent plane to a parametrized surface
C. Area of a parametrized surface

Summary and Focus Questions

A. Surfaces in three-dimensional space can be parameterized in a manner similar to the way curves are parameterized except that we need two parameter variables instead of one. For a domain D, a subset of the uv-plane,

$$\mathbf{r}(u, v) = x(u, v)\mathbf{i} + y(u, v)\mathbf{j} + z(u, v)\mathbf{k}$$

is a surface S in three-dimensional space.

For example, the triangular region D in the uv-plane below is the domain for the parameterization of the surface S (the portion of the plane $z = 6 - 2x - 3y$ in the first octant) using the equation

$$\mathbf{r}(u, v) = (3u)\mathbf{i} + (2v)\mathbf{j} + (6 - 6u - 6v)\mathbf{k}.$$

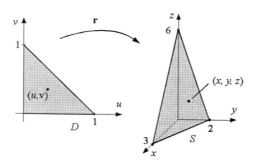

1) Parametrize the cone $z^2 = x^2 + y^2$, $0 \le z \le 1$, using polar coordinates.

Using polar coordinates (r, θ), let D be the unit circle $0 \le \theta \le 2\pi$, $0 \le r \le 1$. Then $x = r \cos \theta$, $y = r \sin \theta$, $z = r$, $0 \le \theta \le 2\pi$, $0 \le r \le 1$ parametrizes the cone.

2) Parametrize the surface of revolution obtained by revolving $y = x^2$, $0 \le x \le 2$ about the x-axis.

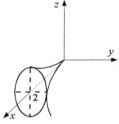

The surface is about the x-axis so we use x as one parameter: $x = x$. For any x, the surface is a circle with radius x^2, so we use θ as the other parameter: $y = x^2 \cos \theta$, $z = x^2 \sin \theta$. Our parametrization is $x = x$, $y = x^2 \cos \theta$, $z = x^2 \sin \theta$ with $0 \le x \le 2$, $0 \le \theta \le 2\pi$.

B. The tangent plane at (x_0, y_0, z_0) to a surface S parametrized by $\mathbf{r}(u, v)$ has as a normal vector $\mathbf{r}_u \times \mathbf{r}_v$. The equation of the tangent plane is

$$(\mathbf{r}_u \times \mathbf{r}_v) \cdot \langle x - x_0, y - y_0, z - z_0 \rangle = 0.$$

S is <u>smooth</u> when the normal vector is not $\mathbf{0}$.

3) Find the equation of the plane tangent to the surface S at $(u, v) = (1, 1)$ where S is parametrized by
$x = 2u^2 v^2$
$y = u + 2v$
$z = 2u + v$.

$\mathbf{r}(u, v) = x\mathbf{i} + y\mathbf{j} + z\mathbf{k}.$
$\mathbf{r}_u = \frac{\partial x}{\partial u}\mathbf{i} + \frac{\partial y}{\partial u}\mathbf{j} + \frac{\partial z}{\partial u}\mathbf{k}$
$\qquad = 4uv^2\mathbf{i} + 1\mathbf{j} + 2\mathbf{k}.$
$\mathbf{r}_v = 4u^2 v\mathbf{i} + 2\mathbf{j} + 1\mathbf{k}.$

$$\mathbf{r}_u \times \mathbf{r}_v = \begin{vmatrix} \mathbf{i} & \mathbf{j} & \mathbf{k} \\ 4uv^2 & 1 & 2 \\ 4u^2v & 2 & 1 \end{vmatrix}$$
$$= -3\mathbf{i} + (8u^2v - 4uv^2)\mathbf{j}$$
$$+ (8uv^2 - 4u^2v)\mathbf{k}.$$

At $(u, v) = (1, 1)$,

$\mathbf{r}_u \times \mathbf{r}_v = -3\mathbf{i} + 4\mathbf{j} + 4\mathbf{k}$

and $(x, y, z) = (2, 3, 3)$.

Thus the tangent plane is

$-3(x - 2) + 4(y - 3) + 4(z - 3) = 0$,

or $3x - 4y - 4z + 18 = 0$.

C. The surface area of a surface S parametrized by $\mathbf{r}(u, v)$ is

$$A(S) = \iint\limits_D |\mathbf{r}_u \times \mathbf{r}_v| dA = \iint\limits_D \sqrt{\left[\frac{\partial(x, y)}{\partial(u, v)}\right]^2 + \left[\frac{\partial(y, z)}{\partial(u, v)}\right]^2 + \left[\frac{\partial(x, z)}{\partial(u, v)}\right]^2} \, dA.$$

3) Set up an iterated integral for the area of each surface S.

a) S is parameterized by

$x = 2uv$

$y = u + v$

$z = u - v$

$0 \le u \le 2, 0 \le v \le 1$.

Let D be the region $0 \le u \le 2, 0 \le v \le 1$.

$\frac{\partial(x, y)}{\partial(u, v)} = \frac{\partial x}{\partial u}\frac{\partial y}{\partial v} - \frac{\partial x}{\partial v}\frac{\partial y}{\partial u}$

$\qquad = (2v)(1) - (2u)(1) = 2v - 2u.$

$\frac{\partial(x, z)}{\partial(u, v)} = (2v)(-1) - (2u)(1) = -2v - 2u.$

$\frac{\partial(y, z)}{\partial(u, v)} = (1)(-1) - (1)(1) = -2.$

The area is

$\int_0^2 \int_0^1 \sqrt{(2v - 2u)^2 + (-2v - 2u)^2 + (-2)^2} \, dv \, du$

$\qquad = \int_0^2 \int_0^1 2\sqrt{2u^2 + 2v^2 + 1} \, dv \, du.$

b) S is the surface
$z = x^2 - y^2$,
$0 \le x \le 1, 0 \le y \le 1$.

Parametrize S with $x = u$, $y = v$,
$z = u^2 - v^2$.
$\frac{\partial(x,y)}{\partial(u,v)} = 1(1) - (0)0 = 1$

$\frac{\partial(x,z)}{\partial(u,v)} = 1(-2v) - 1(2u) = -2v$

$\frac{\partial(y,z)}{\partial(u,v)} = 0(-2v) - 1(2u) = -2u$

Since $x = u$ and $y = v$ we have
$0 \le u \le 1, 0 \le v \le 1$.
The area is
$\int_0^1 \int_0^1 \sqrt{1^2 + (-2v^2) + (-2u)^2} \, dv \, du$

$= \int_0^1 \int_0^1 \sqrt{1 + 4v^2 + 4u^2} \, dv \, du.$

Surface Integrals

Concepts to Master

A. Surface integrals
B. Orientation of a surface; Surface integral of a vector field

Summary and Focus Questions

A. A surface integral over a surface S is a generalization of a line integral over a curve.

Suppose S is a surface parametrized by $\mathbf{r}(u, v) = x(u, v)\mathbf{i} + y(u, v)\mathbf{j} + z(u, v)\mathbf{k}$ for (u, v) in a region D. The <u>surface integral</u> of a function $f(x, y, z)$ whose domain includes S is

$$\iint_S f(x, y, z)\, dS = \iint_D f(\mathbf{r}(u, v))|\mathbf{r}_u \times \mathbf{r}_v|\, dA.$$

In the special case where S is parametrized by a function $z = g(x, y)$,

$$\iint_S f(x, y, z)\, dS = \iint_D f(x, y, g(x, y))\sqrt{(g_x)^2 + (g_y)^2 + 1}\; dA.$$

1) Find an iterated integral for the surface integral of $f(x, y, z) = x$ and S is that portion of the plane $6x + 3y + 2z = 12$ in the first octant.

A parametrization of S is
$\mathbf{r}(u, v) = u\mathbf{i} + v\mathbf{j} + \left(6 - 3u - \frac{3}{2}v\right)\mathbf{k}$,
for $(u, v) \in D$,
$0 \le u \le 2,\ 0 \le v \le 4 - 2u$.
See the figure.

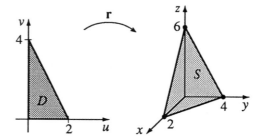

164

$$\mathbf{r}_u \times \mathbf{r}_v = \begin{vmatrix} \mathbf{i} & \mathbf{j} & \mathbf{k} \\ 1 & 0 & -3 \\ 0 & 1 & -\frac{3}{2} \end{vmatrix} = 3\mathbf{i} + \tfrac{3}{2}\mathbf{j} + \mathbf{k}.$$

$$|\mathbf{r}_u \times \mathbf{r}_v| = \sqrt{3^2 + \left(\tfrac{3}{2}\right)^2 + 1^2} = \tfrac{7}{2}.$$

$$\iint_S x\, dS = \iint_D u\left(\tfrac{7}{2}\right) dA = \int_0^2 \int_0^{4-2u} \tfrac{7u}{2}\, dv\, du.$$

B. If it is possible to choose a unit normal vector **n** for each point of a surface S, then S is <u>oriented</u> with orientation provided by the given choice of **n**. For a closed surface (the boundary of a solid) a <u>positive orientation</u> points outward from the solid.

If $\mathbf{F} = P\mathbf{i} + Q\mathbf{j} + R\mathbf{k}$ is a continuous vector field whose domain includes an oriented surface S (with unit normal **n**), the <u>surface integral</u> or <u>flux of **F**</u> over S is defined as

$$\iint_S \mathbf{F} \cdot d\mathbf{S} = \iint_S \mathbf{F} \cdot \mathbf{n}\, dS.$$

When S is given by $\mathbf{r}(u, v) = x(u, v)\mathbf{i} + y(u, v)\mathbf{j} + z(u, v)\mathbf{k}$, for (u, v) in a domain D

$$\iint_S \mathbf{F} \cdot d\mathbf{S} = \iint_D \mathbf{F} \cdot (\mathbf{r}_u \times \mathbf{r}_v)\, dA$$

and in the special case where S is given by $z = g(x, y)$,

$$\iint_S \mathbf{F} \cdot d\mathbf{S} = \iint_D \left(-P\frac{\partial g}{\partial x} - Q\frac{\partial g}{\partial y} + R\right) dA.$$

2) Find an iterated integral for $\iint_S \mathbf{F} \cdot d\mathbf{S}$

where S is the top half of the sphere $x^2 + y^2 + z^2 = 36$ and $\mathbf{F}(x, y, z) = z\mathbf{k}$.

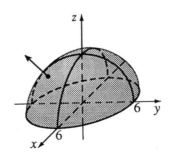

In spherical coordinates, the hemisphere is $\rho = 6$ with $0 \leq \phi \leq \frac{\pi}{2}$.

A parametrization is $x = 6 \sin \phi \cos \theta$, $y = 6 \sin \phi \sin \theta$, $z = 6 \cos \phi$, $0 \leq \phi \leq \frac{\pi}{2}$, $0 \leq \theta \leq 2\pi$.

$\mathbf{r}_\phi \times \mathbf{r}_\theta$

$$= \begin{vmatrix} \mathbf{i} & \mathbf{j} & \mathbf{k} \\ 6 \cos \phi \cos \theta & 6 \cos \phi \sin \theta & -6 \sin \phi \\ -6 \sin \phi \sin \theta & 6 \sin \phi \cos \theta & 0 \end{vmatrix}$$

$$= 36(\sin^2 \phi \cos \theta \mathbf{i} + \sin^2 \phi \sin \theta \mathbf{j} + \sin \phi \cos \theta \mathbf{k}).$$

$\mathbf{F} \cdot (\mathbf{r}_\phi \times \mathbf{r}_\theta) = z(36 \sin \phi \cos \phi)$

$= 6 \cos \phi \, 36 \sin \phi \cos \phi = 216 \sin \phi \cos^2 \phi.$

$\iint_S \mathbf{F} \cdot d\mathbf{S} = \int_0^{\pi/2} \int_0^{2\pi} 216 \sin \phi \cos^2 \phi \; d\theta \; d\phi.$

Stokes' Theorem

Concepts to Master

Stokes' Theorem

Summary and Focus Questions

Stokes' Theorem is a version of Green's Theorem for three dimensions -- relating a surface integral over a surface S to a line integral around the boundary of S.

Stokes' Theorem: Let S be a smooth, simply connected, orientable surface bounded by a simple closed curve C. Let $\mathbf{F} = P\mathbf{i} + Q\mathbf{j} + R\mathbf{k}$, where P, Q, R have continuous first partials. Finally let C have a positive orientation (looking down from any outer unit normal to the surface positive orientation means counterclockwise). Then

$$\int_C \mathbf{F} \cdot d\mathbf{r} = \iint_S (\text{curl } \mathbf{F}) \cdot d\mathbf{S}.$$

1) Use Stokes' Theorem to rewrite $\int_C \mathbf{F} \cdot d\mathbf{r}$ where $\mathbf{F}(x, y, z) = xy\mathbf{i} + yz\mathbf{j} + xz\mathbf{k}$ and C is the (counterclockwise) oriented triangular boundary of the intersection of the plane $x + 2y + 4z = 8$ with the first octant.

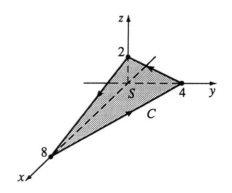

Let S be a plane surface bounded by C.
By Stokes' Theorem:
$\int_C \mathbf{F} \cdot d\mathbf{r} = \iint_S \text{curl } \mathbf{F} \cdot d\mathbf{S}$.
curl $\mathbf{F} = (0 - y)\mathbf{i} + (0 - z)\mathbf{j} + (0 - x)\mathbf{k}$
$\qquad = -y\mathbf{i} - z\mathbf{j} - x\mathbf{k}$.
The surface S is
$x = 8 - 2u - 4v$, $y = u$, $z = v$
$0 \le u \le 4$, $0 \le v \le 2 - \frac{u}{2}$.
$$\mathbf{r}_u \times \mathbf{r}_v = \begin{vmatrix} \mathbf{i} & \mathbf{j} & \mathbf{k} \\ -2 & 1 & 0 \\ -4 & 0 & 1 \end{vmatrix} = \mathbf{i} + 2\mathbf{j} + 4\mathbf{k}$$
is the normal vector to the plane.
curl $\mathbf{F} \cdot (\mathbf{r}_u \times \mathbf{r}_v)$
$\qquad = -y(1) + (-z)2 + (-x)4$
$\qquad = -y - 2z - 4x$
$\qquad = -u - 2v - 4(8 - 2u - 4v)$
$\qquad = 7u + 14v - 32$.
Thus $\iint_S \text{curl } \mathbf{F} \cdot d\mathbf{S}$
$\qquad = \int_0^4 \int_0^{2-(u/2)} (7u + 14v - 32)\, dv\, du$.

2) Use Stokes' Theorem to rewrite $\iint_S \text{curl } \mathbf{F} \cdot d\mathbf{S}$, where
$$\mathbf{F}(x,\, y,\, z) = (x - y)\mathbf{i} + (y - z)\mathbf{j} + (x - z)\mathbf{k}$$
and S is the paraboloid $z = 4 - x^2 - y^2$, $z \ge 0$, oriented upward.

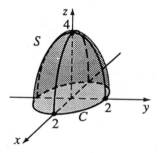

The boundary of S is the circle C in the xy plane $x^2 + y^2 = 4$, $z = 0$.
$\mathbf{r}(t) = 2\cos t\mathbf{i} + 2\sin t\mathbf{j}$, $0 \le t \le 2\pi$.
$\mathbf{r}'(t) = -2\sin t\mathbf{i} + 2\cos t\mathbf{j}$.
$\mathbf{F}(\mathbf{r}(t)) = 2((\cos t - \sin t)\mathbf{i} + \sin t\mathbf{j} + \cos t\mathbf{k})$
$\mathbf{F}(\mathbf{r}(t)) \cdot \mathbf{r}'(t) = 4(-\cos t\sin t + \sin^2 t$
$\qquad\qquad\qquad + \cos t\sin t)$
$\qquad = 4\sin^2 t$.
Thus $\iint_S \text{curl } \mathbf{F} \cdot d\mathbf{S} = \iint_C \text{curl } \mathbf{F} \cdot d\mathbf{r}$
$\qquad = \int_0^{2\pi} 4\sin^2 t\, dt$.

The Divergence Theorem

Concepts to Master

Divergence Theorem

Summary and Focus Questions

<u>The Divergence Theorem</u> is an extension of Green's Theorem to three dimensions, relating surface integrals and triple integrals. Let E be a simple solid whose boundary surface S has positive orientation. If $\mathbf{F} = P\mathbf{i} + Q\mathbf{j} + R\mathbf{k}$ with P, Q, and R having continuous partial derivatives on an open region containing E, then

$$\iint\limits_{S} \mathbf{F} \cdot d\mathbf{S} = \iiint\limits_{E} \operatorname{div} \mathbf{F} \, dV.$$

1) Write a triple iterated integral for $\iint\limits_{S} \mathbf{F} \cdot d\mathbf{S}$, where

$\mathbf{F}(x, y, z) = xy\mathbf{i} + y^2\mathbf{k} + xz\mathbf{k}$ and S is the triangular surface bounded by $6x + 4y + 3z = 12$ and the first octant.

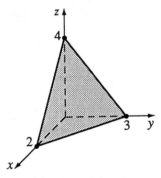

Let E be the solid region enclosed by S:

$0 \leq x \leq 2$

$0 \leq y \leq 3 - \frac{3}{2}x$

$0 \leq z \leq 4 - 2x - \frac{4}{3}y$

$\operatorname{div} \mathbf{F} = \frac{\partial}{\partial x}(xy) + \frac{\partial}{\partial y}(y^2) + \frac{\partial}{\partial z}(xz)$

$\quad = y + 2y + x = x + 3y.$

Thus $\iint\limits_{S} \mathbf{F} \cdot d\mathbf{S} = \iiint\limits_{E}(x + 3y)dV$

$= \int_0^2 \int_0^{3-(3/2)x} \int_0^{4-2x-(4/3)y}(x + 3y)dz \, dy \, dx.$

15

Differential Equations

"DOES THIS APPLY ALWAYS, SOMETIMES, OR NEVER?"

Cartoons courtesy of Sidney Harris. Used by permission.

Basic Concepts; Separable & Homogeneous Equations

Concepts to Master

A. Ordinary differential equations; Order; Separable equations; Solution curves; Direction fields

B. Initial conditions

C. Euler's Method

D. Homogeneous differential equations

Summary and Focus Questions

A. An <u>ordinary differential equation</u> is an equation involving x, y, $\frac{dy}{dx}$, $\frac{d^2y}{dx^2}$, $\cdots$, $\frac{d^n y}{dx^n}$. The n in the highest $\frac{d^n y}{dx^n}$ that appears in the equation is the <u>order</u> of the equation. A <u>first order</u> equation has the form $\frac{dy}{dx} = F(x, y)$.

A <u>separable</u> differential equation has the form $\frac{dy}{dx} = \frac{g(x)}{h(y)}$ which can be solved by integrating $\int h(y)dy = \int g(x)dx$ and solving for y.

A <u>particular solution</u> to a differential equation is a function $y = f(x)$ that satisfies the equation. Its graph is called a <u>solution curve</u>. A <u>general solution</u> is an expression with arbitrary constants to represent all particular solutions.

Solution curves to $\frac{dy}{dx} = F(x, y)$ may be visualized by drawing a <u>direction field</u> -- short line segments at various points (x, y) with slope $F(x, y)$.

<u>Example:</u>
The separable equation $y' = \frac{1}{3y^2}$ has general solution
$$\int 3y^2 \, dy = \int 1 \, dx$$
$$y^3 = x + C$$
$$y = \sqrt[3]{x + C}.$$

Some particular solutions are $y = \sqrt[3]{x+1}$, $y = \sqrt[3]{x-4}$, and so on. The direction field is pictured below:

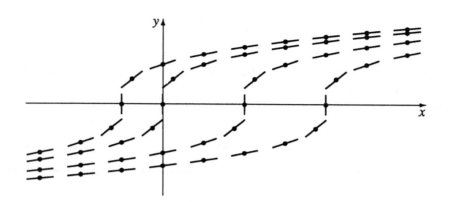

1) The equation $x^4 \frac{d^2y}{dx^2} + 5xy^2 \frac{dy}{dx} = x + y$
   has order _____.

2) Is $\frac{dy}{dx} = 2xy$ separable?

3) Sketch a direction field for $\frac{dy}{dx} = 2xy$.

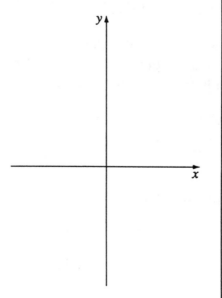

2.

Yes. $\frac{dy}{dx} = 2xy$ may be rewritten as
$\frac{dy}{y} = 2x \ dx$.

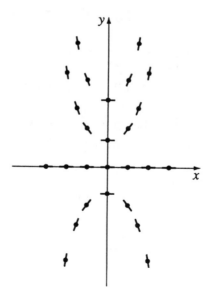

4) Solve $\frac{dy}{dx} = 2xy$ and draw three solution curves. (Compare your curves to the direction field in question 3.)

$\frac{dy}{dx} = 2xy$, so $\frac{dy}{y} = 2x \, dx$

$\int \frac{dy}{y} = \int 2x \, dx$.

$\ln|y| = x^2 + C$

$|y| = e^{x^2+C} = e^{x^2} e^C$.

$y = Ke^{x^2}$, K a real number.

Solution curves for $K = 1$, 2, and -1 are given in the following figure.

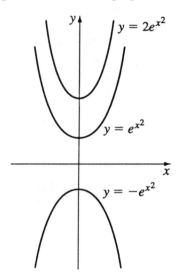

B. An <u>initial condition</u> for a differential equation is a set of values (x_0, y_0) for the variables that determine a particular solution from the general solution. After finding the general solution which includes a constant of integration, C, use the values (x_0, y_0) to find a value for C.

5) Solve $x\frac{dy}{dx} = -y^2 (x > 0)$ with the initial condition of $x_0 = 1$, $y_0 = \frac{1}{3}$.

Separate variables: $-y^{-2} \, dy = x^{-1} \, dx$.

Integrate: $\int -y^{-2} \, dy = \int x^{-1} \, dx$.

$\frac{1}{y} = \ln|x| + C = \ln x + C$.

$y = \frac{1}{\ln x + C}$ is the general solution.

For $x_0 = 1$, $y_0 = \frac{1}{3}$, $\frac{1}{3} = \frac{1}{\ln 1 + C} = \frac{1}{C}$, so $C = 3$. The particular solution is $y = \frac{1}{\ln x + 3}$.

C. Euler's method is a way to find approximations to initial value problems based on following tangent lines in direction fields.

For a differential equation $\frac{dy}{dx} = F(x, y)$ with initial condition (x_0, y_0) Euler's method constructs a table of (x, y) values that estimate the solution. The x_i values are evenly spaced out (a <u>step size</u> of h between successive x_i). The y_{i+1} is calculated from the direction field at (x_i, y_i):

x	y
x_0	y_0 (given)
$x_1 = x_0 + h$	$y_1 = y_0 + hF(x_0, y_0)$
$x_2 = x_1 + h$	$y_2 = y_1 + hF(x_1, y_1)$
$x_3 = x_2 + h$	$y_3 = y_2 + hF(x_2, y_2)$
$\vdots$	
x_{n+1}	$y_{n+1} = y_n + hF(x_n, y_n)$

As you step further from x_0, the y_i become less accurate, but this can be partially overcome by choosing h to be small.

6) Approximate five solution values to
$\frac{dy}{dx} = 2xy$, $y(1) = 3$ with step size 0.2.

x	y
$x_0 = 1$	$y_0 = 3$
$x_1 = 1.2$	$y_1 = 3 + (0.2)(2(1)(3)) = 4.2$
$x_2 = 1.4$	$y_2 = 4.2 + (0.2)(2(1.2)(4.2))$
	$= 6.216$
$x_3 = 1.6$	$y_3 = 9.696$
$x_4 = 1.8$	$y_4 = 15.903$
$x_5 = 2.0$	$y_5 = 27.353$

D. The equation $y' = f(x, y)$ is <u>homogeneous</u> if $f(x, y)$ can be written as a single variable function $g(v)$ with argument $v = \frac{y}{x}$. For example,

$y' = \frac{x^2 + y^2}{x^2 - y^2}$ is homogeneous because it can be written as

$y' = \frac{1 + \left(\frac{y}{x}\right)^2}{1 - \left(\frac{y}{x}\right)^2}$, which is $g(v) = \frac{1 + v^2}{1 - v^2}$ for $v = \frac{y}{x}$.

To solve a homogeneous equation, use the change of variable $v = \frac{y}{x}$, which transforms the equation into the separable equation $v + xv' = g(v)$. The solution to the original equation is $y = xv(x)$.

6) Is $y' = \frac{x^2+2xy+y^2}{3x^2+y^2}$ homogeneous?

Yes. Dividing numerator and denominator by x^2 yields
$$y' = \frac{1+2\left(\frac{y}{x}\right)+\left(\frac{y}{x}\right)^2}{3+\left(\frac{y}{x}\right)^2}.$$

7) Solve $y' = \frac{x^2+y^2}{2xy}$.

$$y' = \frac{x^2+y^2}{2xy} = \frac{1+\left(\frac{y}{x}\right)^2}{2\left(\frac{y}{x}\right)}.$$
Let $v = \frac{y}{x}$, $g(v) = \frac{1+v^2}{2v} = \frac{1}{2v} + v$.
Then $v + xv' = g(v)$ is $v + xv' = \frac{1}{2v} + v$.
$v' = \frac{1}{2vx}$. Therefore $2vv' = \frac{1}{x}$.
Thus $\int 2v \, dv = \int \frac{1}{x} \, dx$.
$v^2 = \ln|x| + C$
$v = \pm \sqrt{\ln|x| + C}$.
Thus $y = xv = \pm x\sqrt{\ln|x| + C}$.

First Order Linear Equations

Concepts to Master

Linear differential equations; Solutions by integrating factors

Summary and Focus Questions

A first order differential equation is <u>linear</u> if it can be written as

$$y' + P(x)y = Q(x)$$

where $P(x)$ and $Q(x)$ are continuous. The equation is solved by multiplying both sides by the <u>integrating factor</u> $I(x) = e^{\int P(x)dx}$ and solving $Iy = \int IQ\ dx$ for y.

1) Is $y' + \frac{x}{y} = x^2 + 1$ linear?

    No, because of the term $\frac{x}{y}$.

2) What integrating factor should be used for $y' + 7y = x^2 + x$?

    $I(x) = e^{\int 7\ dx} = e^{7x}$.

3) Find the general solution to $xy' + y = e^{-x}$, $x > 0$.

    Put in standard form: $y' + \frac{1}{x}y = \frac{e^{-x}}{x}$.

    $I(x) = e^{\int 1/x\ dx} = e^{\ln x} = x$.

    $xy = \int x\frac{e^{-x}}{x}\ dx = \int e^{-x}\ dx$

        $= -e^{-x} + C$.

    Thus $y = \frac{-e^{-x}+C}{x}$.

Exact Equations

Concepts to Master

A. Exact differential equations
B. Integrating factors

Summary and Focus Questions

A. A differential equation of the form $P(x, y) + Q(x, y)y' = 0$ is <u>exact</u> means that there is a function $f(x, y)$ such that $P = f_x$ and $Q = f_y$. Equivalently, the differential equation is exact if and only if $P_y = Q_x$.

The solution to an exact equation is given implicitly by the equation $f(x, y) = C$.

1) Is the following equation exact?
$(10xy - 3y^2)y' = 2x - 5y^2$

Written in standard $P + Qy' = 0$ form the equation is
$-2x + 5y^2 + (10xy - 3y^2)y' = 0.$
Both $P_y = 10y$ and $Q_x = 10y$, so the equation is exact.

2) Solve the exact equation
$2xye^{x^2} + 1 + e^{x^2}y' = 0.$

Let $P(x, y) = 2xye^{x^2} + 1$ and
$Q(x, y) = e^{x^2}.$ $\int P \, dx = ye^{x^2} + x + g(y).$
Differentiate this with respect to y and compare with $Q(x, y)$:
$e^{x^2} + g'(y) = e^{x^2}$
$g'(y) = 0,$ so let $g(y) = 0.$
Thus the solution is given implicitly by
$f(x, y) = ye^{x^2} + x = C.$

Thus $y = \frac{C-x}{e^{x^2}}.$

B. Sometimes, when $P(x, y) + Q(x, y)y' = 0$ is not exact, there is an
<u>integrating factor</u> $I(x, y)$ such that

$$P(x, y)I(x, y) + Q(x, y)I(x, y)y' = 0 \text{ is exact.}$$

$I(x, y)$, if it exists, is a solution to $PI_y - QI_x = I(Q_x - P_y)$.

In general the equation is difficult to solve, but there are two special cases:

i)   If $\frac{P_y - Q_x}{Q}$ is a function of x, then I is a function of x and $\frac{dI}{dx} = \frac{P_y - Q_x}{Q}I$.

ii) If $\frac{Q_x - P_y}{Q}$ is a function of y, then I is a function of y and $\frac{dI}{dy} = \frac{Q_x - P_y}{P}I$.

3)   Use the integrating factor xy to solve
     $(4y^2 + 3xy) + (6xy + 2x^2)y' = 0$.

> Multiply by xy:
> $(4xy^3 + 3x^2y^2) + (6x^2y^2 + 2x^3y)y' = 0$.
> Let $P(x, y) = 4xy^3 + 3x^2y^2$ and
> $Q(x, y) = 6x^2y^2 + 2x^3y$.
> The equation is exact because
> $P_y = 12xy^2 + 6x^2y = Q_x$.
> Integrate $P(x, y)$:
> $\int(4xy^3 + 3x^2y^2)dx = 2x^2y^3 + x^3y^2 + g(y)$.
> Differentiate with respect to y and compare
> to $Q(x, y)$:
> $6x^2y^2 + 2x^3y + g'(y) = 6x^2y^2 + 2x^3y$.
> $g'(y) = 0$, so let $g(y) = 0$.
> The solution is given implicitly by
> $2x^2y^3 + x^3y^2 = C$.

4)   Find an integrating factor for
     $2xy^2 + 2y + (4x^2y + 6x)y' = 0$.

> Let $P(x, y) = 2xy^2 + 2y$ and
> $Q(x, y) = 4x^2y + 6x$.
> $P_y = 4xy + 2$, $Q_x = 8xy + 6$.
> $\frac{Q_x - P_y}{P} = \frac{(8xy+6)-(4xy+2)}{2xy^2+2y} = \frac{2}{y}$.
> This is a function of y only.
> Next solve the separable equation $\frac{dI}{dy} = \frac{2}{y}I$.
> Separate: $\frac{dI}{I} = \frac{2}{y} dy$ and integrate:
> $\int\frac{dI}{I} = \int\frac{2}{y} dy$. $\ln|I| = 2\ln|y| = \ln y^2$.
> Thus $I(x, y) = y^2$ is an integrating factor.

Strategy for Solving First Order Equations

Concepts to Master

There are no new concepts in this section. Instead we pull together the solution techniques of previous sections.

Summary and Focus Questions

For first order equations the form of the equation suggests which method to apply. Here is a brief summary of each form together with a representative example.

Name/Form	Method of Solution	Example
Separable $\frac{dy}{dx} = g(x)h(y)$	Integrate $\int \frac{dy}{h(y)} = \int g(x)dx$ and solve for y.	$\frac{dy}{dx} = \frac{x^2}{2y+1}$ $\int (2y+1)dy = \int x^2\ dx$
Linear $\frac{dy}{dx} + P(x)y = Q(x)$	Multiply by the integrating factor $I(x) = e^{\int P(x)dx}$ and solve $Iy = \int IQ\ dx$ for y.	$y' + 4xy = 2x + 1$ $P(x) = 4x,\ I(x) = e^{2x^2}$ $e^{2x^2}y = \int e^{2x^2}(2x+1)dx.$
Homogeneous $\frac{dy}{dx} = g\left(\frac{y}{x}\right)$	Use the change of variable $v = \frac{y}{x}$ and solve $v + xv' = g(v)$ for v. Then $y = xv$.	$y' = \frac{2xy^2 + x^2 y}{x^3 + y^3}$ $= \frac{2\left(\frac{y}{x}\right)^2 + \left(\frac{y}{x}\right)}{1 + \left(\frac{y}{x}\right)^3}$
Exact $P(x, y) + Q(x, y)y' = 0$, where $P_y = Q_x$.	Find $f(x, y)$ such that $\nabla f = \langle P, Q \rangle$. The implied solution satisfies $f(x, y) = C$. In case $P_y \neq Q_x$, try to find an integrating factor $I(x, y)$ so that the equation becomes exact.	$4xy^3 + 1 + 6x^2y^2y' = 0$ $P_y = Q_x = 12xy^2$ $f(x, y) = 2x^2y^3 + x = C$

Name the method of solution for each.

1) $xyy' = y^2 + 1$

Separable. $\int \frac{y}{y^2+1}\, dy = \int \frac{dx}{x}$.

2) $3y^3 + 1 + 9xy^2 y' = 0$

Exact. For $P(x,\, y) = 3y^3 + 1$ and $Q(x,\, y) = 9xy^2$, $P_y = Q_x = 9y^2$.

3) $y' + 2y = e^x$

Linear. $I(x) = e^{\int 2\, dx} = e^{2x}$.

4) $x^2 y' + xy = y^2$

Homogeneous. Rewrite as $x^2 y' = y^2 - xy$:

$$y' = \frac{y^2 - xy}{x^2} = \left(\frac{y}{x}\right)^2 - \frac{y}{x}.$$

5) $x^2 y'' + 2xy' = y^3 + e^{xy}$

None of these methods work; this is not a first order equation.

Second Order Linear Equations

Concepts to Master

A. Second order linear equations; Homogeneous; Linear combination; Linearly independent; Auxiliary equation

B. Initial-values; Boundary-values

Summary and Focus Questions

A. A <u>second order linear differential equation</u> has the form

$$P(x)y'' + Q(x)y' + R(x)y = G(x).$$

This section considers the <u>homogeneous</u> case, where $G(x) = 0$.

If y_1 and y_2 are two solutions to the homogeneous equation then all linear combinations $c_1y_1 + c_2y_2$ are also solutions (c_1c_2 real numbers).

Solutions y_1 and y_2 are <u>linearly independent</u> means neither y_1 nor y_2 is a constant multiple of the other.

If y_1 and y_2 are linearly independent solutions, then *all* solutions may be written as linear combinations of y_1 and y_2.

In the case where P, Q, and R are constants, the equation has the form $ay'' + by' + cy = 0$.

The corresponding <u>auxiliary equation</u> (with variable r) is $ar^2 + br + c = 0$. The type of roots (real or imaginary) of the auxiliary equation determine the type of solution to $ay'' + by' + cy = 0$:

Discriminant	Roots	Solution
$b^2 - 4ac > 0$	r_1, r_2 real, $r_1 \neq r_2$	$y = c_1 e^{r_1 x} + c_2 e^{r_2 x}$
$b^2 - 4ac = 0$	r, real	$y = c_1 e^{rx} + c_2 x e^{rx}$
$b^2 - 4ac < 0$	r_1, r_2 complex, $r_1 \neq r_2$	$y = e^{\alpha x}(c_1 \cos \beta x + c_2 \sin \beta x)$
	$r_1 = \alpha + \beta i$	
	$r_2 = \alpha - \beta i$	

1) True or False:
   $xy' + y'' + 5x + 10y = 0$ is
   homogeneous.

   False (because of the term $5x$).

2) True or False:
   $y_1 = 2xy - 3y^2$ and $y_2 = 12y^3 - 8xy$
   are linearly independent.

   False, $-4y_1 = y_2$, so they are dependent.

3) Sometimes, Always, or Never:
   If y_1 and y_2 are solutions to
   $ay'' + by' + cy = 0$ then all solutions
   are of the form $c_1 y_1 + c_2 y_2$.

   Sometimes; y_2 and y_1 must be linearly
   independent for this to be true.

4) Find the general solution to each:

   a) $y'' + 6y' + 8y = 0$

   The auxiliary equation is $r^2 + 6r + 8 = 0$.
   $(r + 2)(r + 4) = 0$.
   Thus $r = -2, -4$.
   The general solution is
   $y = c_1 e^{-2x} + c_2 e^{-4x}$.

   b) $y'' - 6y' + 13y = 0$

   The auxiliary equation is $r^2 - 6r + 13 = 0$.
   $r = \dfrac{-(-6) \pm \sqrt{36 - 4(13)}}{2}$

   $= \dfrac{6 \pm \sqrt{-16}}{2} = 3 \pm 2i$.
   The general solution is
   $y = e^{3x}(c_1 \cos 2x + c_2 \sin 2x)$.

c) $4y'' - 12y' + 9y = 0$

> The auxiliary equation is
> $4r^2 - 12r + 9 = 0$
> $(2r - 3)^2 = 0$
> $r = \frac{3}{2}$.
> The general solution is
> $y = c_1 e^{(3/2)x} + c_2 x e^{(3/2)x}$.

B. To specify a particular solution of a second order equation two conditions must be given. Depending on what is specified we have two classes of problems.

1) If $y(x_0) = y_0$ and $y'(x_0) = y_1$ are specified, we have an <u>initial-value problem</u>.

2) If $y(x_0) = y_0$ and $y(x_1) = y_1$ are specified, we have a <u>boundary-value problem</u>.

For continuous P, Q, R, and G with $P \neq 0$, initial-value problems will have a solution but boundary-value problems may or may not. The method of solution involves substituting the conditions in y and/or y' and solving two equations in the two unknowns c_1 and c_2.

5)   Solve $y'' + 2y' - 8y = 0$ with initial conditions $y(0) = 28$ and $y'(0) = 2$.

> $r^2 + 2r - 8 = 0$ has solutions $r = -4, 2$.
> The general solution is
> $y = c_1 e^{2x} + c_2 e^{-4x}$.
> Since $y(0) = 28$, $c_1 + c_2 = 28$.
> $y' = 2c_1 e^{2x} - 4c_2 e^{-4x}$.
> Since $y'(0) = 2$, $2c_1 - 4c_2 = 2$
> or $c_1 - 2c_2 = 1$
> The system $c_1 + c_2 = 28$
> $c_1 - 2c_2 = 1$
> has solution $3c_2 = 27$
> $c_2 = 9$
> Thus $c_1 - 18 = 1$, $c_1 = 19$.
> The particular solution is
> $y = 19e^{2x} + 9e^{-4x}$.

6) Solve $y'' - 4y' + 3y = 0$ with boundary conditions $y(0) = e^2$, $y(1) = e$.

$$r^2 - 4r + 3 = 0$$
$$(r - 3)(r - 1) = 0$$
$$r = 1, 3.$$
The general solution is
$y = c_1 e^x + c_2 e^{3x}$.
$y(0) = e^2$ implies $c_1 + c_2 = e^2$.
$y(1) = e$ implies $c_1 e + c_2 e^3 = e$,
or $c_1 + c_2 e^2 = 1$.
Subtract the second equation from the first:
$c_2 - c_2 e^2 = 1 - e^2$
$c_2(1 - e^2) = 1 - e^2$
$c_2 = 1.$
Then $c_1 + 1 = e^2$ and so $c_1 = 1 - e^2$.
The solution is $y = (1 - e^2)e^x + e^{3x}$.

7) Solve $y'' - 6y' + 9y = 0$ with boundary conditions $y(0) = 2$, $y(1) = e^3$.

$r^2 - 6r + 9 = 0$ has solution $r = 3$.
The general solution is
$y = c_1 e^{3x} + c_2 x e^{3x}$.
$y(0) = 2$ implies $c_1 = 2$
$y(1) = 1$ implies $c_1 e^3 + c_2 e^3 = e^3$
$c_1 + c_2 = 1$. Since $c_1 = 2$, $c_2 = -1$.
Thus $y = 2e^{3x} - x e^{3x}$.

Nonhomogeneous Linear Equations

Concepts to Master

A. Nonhomogeneous equation; Solution by the Method of Undetermined Coefficients

B. Solution by the Method of Variation of Parameters

Summary and Focus Questions

A. A second order <u>linear nonhomogeneous differential equation</u> with constant differential coefficients has the form $ay'' + by' + cy = G(x)$, where $G(x)$ is continuous. The corresponding <u>complementary equation</u> is $ay'' + by' + cy = 0$.

The general solution is $y(x) = y_p(x) + y_c(x)$ where y_c is the general solution of the complementary equation and y_p is a particular solution of the nonhomogeneous equation.

The particular solution y_p can sometimes be found using the <u>method of undetermined coefficients</u>. The solution $y_p(x)$ is given in the following table and depends on the form of $G(x)$.

$G(x)$ **form**	**Possible** $y_p(x)$ **form**
polynomial: $C_n x^n + \cdots + C_1 x + C_0$	$A_x x^n + \cdots + A_1 x + A_0$
exponential: Ce^{kx}	$A_1 x^r e^{kx}$, where $r = 0$, 1, or 2
trigonometric: $C \cos kx + D \sin kx$	$A_1 x^r \cos kx + A_2 x^r \sin kx$, where $r = 0$ or 1.

Note: In the last two cases r is the smallest nonnegative integer such that $y_p(x)$ is *not* a solution to the complementary equation First try $r = 0$. If any term of $y_p(x)$ is a solution to the complementary equation, then try $r = 1$ or if necessary, $r = 2$.

We then substitute $y_p(x)$ and its derivatives into the nonhomogeneous equation to obtain a system of equations to determine A_1, A_2, ...

185

1) Solve the equation
$$y'' + 3y' - 10y = 16e^{3x}.$$

The complementary equation
$y'' + 3y' - 10y = 0$ has auxiliary equation
$r^2 + 3r - 10 = 0$. $(r - 2)(r + 5) = 0$, so
$r = 2, -5$. The general solution to the
nonhomogeneous equation is
$y(x) = c_1 e^{2x} + c_2 e^{-5x} + y_p(x)$. To find
$y_p(x)$ we note the form of $G(x)$ is $16e^{3x}$ so
$y_p(x) = Ax^r e^{3x}$ is a good trial form. First
consider $r = 0$, $y_p(x) = Ae^{3x}$.
$y_p'(x) = 3Ae^{3x}$, $y_p''(x) = 9Ae^{3x}$.
Substituting these into the original equation
yields
$9Ae^x + 3(3Ae^{2x}) - 10(Ae^{3x}) = 16e^{2x}$
$8Ae^{3x} = 16e^{2x}$, $A = 2$.
Hence $y_p(x) = 2e^{3x}$. (We note $2e^{3x}$ does
not satisfy the complementary equation so
$r = 0$ is valid.) The general solution is
$y(x) = c_1 e^{2x} + c_2 e^{-5x} + 2e^{3x}$.

2) Find a particular solution to
$$y'' - 7y' + 3y = 6x - 2.$$

The form $6x - 2$ suggests
$y_p(x) = Ax + B$. $y_p'(x) = A$, $y_p''(x) = 0$.
Substituting into the original equation:
$0 - 7(A) + 3(Ax + B) = 6x - 2$
$3Ax + (-7A + 3B) = 6x - 2$
Comparing coefficients
$3A = 6$ and $-7A + 3B = -2$.
Thus $A = 2$ and $-7(2) + 3B = -2$,
$B = 4$. Therefore $y_p(x) = 2x + 4$.

B. The method of <u>variation of parameters</u> given below, unlike the undetermined
coefficients method, does not rely on the form of $G(x)$ to find a particular
solution $y_p(x)$ to $ay'' + by' + cy = G(x)$:

1) Find linearly independent solutions y_1 and y_2 to $ay'' + by' + cy = 0$.
2) Solve the following system for u_1' and u_2':
$$u_1' y_1 + u_2' y_2 = 0$$
$$a(u_1' y_1' + u_2' y_2') = G(x).$$

3) Integrate u_1' and u_2' to get $u_1(x)$ and $u_2(x)$.
4) A particular solution is $y_p(x) = u_1(x)y_1(x) + u_2(x)y_2(x)$.

3) Solve $y'' - 5y' + 4y = 30e^{6x}$.

1) The auxiliary equation is
$y'' - 5y' + 4y = 0$.
From $r^2 - 5r + 4 = 0$,
$(r-1)(r-4) = 0$, $r = 1$, $r = 4$.
Thus the general solution to
$y'' - 5y' + 4y = 0$ is
$y_c(x) = c_1e^x + c_2e^{4x}$.
Let $c_1 = 1$, $c_2 = 0$, then $y_1 = e^x$.
Let $c_1 = 0$, $c_2 = 1$, then $y_2 = e^{4x}$.
y_1 and y_2 are linearly independent and
$y_1' = e^x$, $y_2' = 4e^{4x}$.

2) We solve the system
$u_1'e^x + u_2'e^{4x} = 0$
$u_1'e^x + u_2'(4e^{4x}) = 30e^{6x}$.
From the first equation, $u_1' = -u_2'e^{3x}$.
Substituting into the second,
$(-u_2'e^{3x})e^x + u_2'4e^{4x} = 30e^{6x}$
$3u_2'e^{4x} = 30e^{6x}$
$u_2' = 10e^{2x}$.
Thus $u_1' = -(10e^{2x})e^{3x} = -10e^{5x}$.

3) From $u_1' = -10e^{5x}$, $u_1 = -2e^{5x}$ and
from $u_2' = 10e^{2x}$, $u_2 = 5e^{2x}$.

4) The particular solution is
$y_p(x) = u_1y_1 + u_2y_2$
$= (-2e^{5x})e^x + (5e^{2x})e^{4x} = 3e^{6x}$.
Therefore the general solution to the
nonhomogeneous equation is
$y(x) = c_1e^x + c_2e^{4x} + 3e^{6x}$.

Applications of Second Order Differential Equations

Concepts to Master

Second order equations as models for circuits with resistors, inductors, or capacitors and for damped motion

Summary and Focus Questions

The equation $ay'' + by' + cy = G(x)$ has applications in several areas including

1) Damped motion:
   $mx'' + cx' + kx = F(t)$ where
   $m =$ mass of an object at the end of a spring
   $x =$ displacement from equilibrium at time t
   x', $x'' =$ velocity and acceleration
   $c =$ damping constant (such as air resistance)
   $k =$ spring constant
   $F(t) =$ external force (such as gravity)

2) Electric circuits:
   $LQ'' + RQ' + \frac{1}{c}Q = E(t)$ where
   $Q =$ charge on the capacitor at time t
   $Q' =$ current
   $L =$ inductance constant for an inductor
   $R =$ resistance constant for a resistor
   $\frac{1}{c} =$ elastance constant for a capacitor
   $E(t) =$ electromotive force applied
           to the circuit

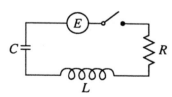

1) A spring with a 10-kg mass can be stretched 0.5 m beyond its equilibrium by a force of 85 N (newtons). Suppose the spring is held in a fluid with damping constant 20. If the mass starts at equilibrium with an initial velocity of 1 m/sec, find the position of the mass after t seconds.

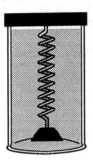

To find the spring constant, k, use Hooke's law: $k(0.5) = 85$, $k = 170$.
The differential equation model is
$10x'' + 20x' + 170x = 0$, or
$x'' + 2x' + 17x = 0$
with initial values $x(0) = 0$, $x'(0) = 1$.
From $r^2 + 2r + 17 = 0$, $r = -1 \pm 4i$.
Thus the general solution is
$x = e^{-t}(c_1 \cos 4t + c_2 \sin 4t)$.
At $t = 0$, $x = 0$, so
$0 = 1(c_1 \cos 0 + c_2 \sin 0)$
Therefore $c_1 = 0$.
Thus $x = e^{-t}(c_2 \sin 4t)$.
$x' = e^{-t}(4c_2 \cos 4t) - e^{-t}(c_2 \sin 4t)$.
At $t = 0$, $x' = 1$, so
$1 = 1(4c_2) - 1(c_2(0))$, $c_2 = 0.25$.
Thus $x = 0.25e^{-t} \sin 4t$ is the position after t seconds.

2) A series circuit consists of a resistor
with $R = 40\Omega$ (ohms), an inductor
with $L = 1$ H (henries), a capacitor
with $C = 0.01$ F (farads), and a
generator producing a voltage of
$3 + 2\sin t$. Find the differential
equation for determining the charge at
time t.

$L = 1,\ R = 40,\ C = 0.01,$
$E(t) = 3 + 2\sin t.$
The equation is
$Q'' + 40Q' + 100Q = 3 + 2\sin t.$

Series Solutions

Concepts to Master

Power series solution to a differential equation; Recursion relation

Summary and Focus Questions

The method of solving a differential equation using a power series has these steps:

1) Assume the equation has a solution of the form
$$y = \sum_{n=0}^{\infty} a_n x^n = a_0 + a_1 x + a_2 x^2 + a_3 x^3 + \ldots$$

2) Obtain expressions for y', y'', etc.
$$y' = \sum_{n=1}^{\infty} n a_n x^{n-1} = a_1 + a_2 x + a_3 x^2 + \ldots$$

$$y'' = \sum_{n=2}^{\infty} n(n-1) a_n x^n = a_2 + a_3 x + a_4 x^2 + \ldots$$

3) Substitute these expressions into the differential equation.

4) Equate coefficients of corresponding powers of x to obtain equations to determine a_0, a_1, a_2, ...

A general formula that expresses a_i in terms of previous a_j, where $j < i$, is called a <u>recursion relation</u>. (These frequently occur in step 4.)

1) Find a series solution to $y'' = xy'$.

1) $y = a_0 + a_1 x + a_2 x^2 + a_3 x^3 + \ldots$,
2) $y' = a_1 + 2a_2 x + 3a_3 x^2 + 4a_4 x^3 + \ldots$, and
$y'' = 2a_2 + 3 \cdot 2a_3 x + 4 \cdot 3a_4 x^2 + 5 \cdot 4a_5 x^3 + \ldots$.

Thus $xy' = a_1 x + 2a_2 x^2 + 3a_3 x^3 + \dots$.
From the equation $y'' = xy'$ we equate coefficients.

Power of x	Relation
x^0	$2a_2 = 0$
x^1	$3 \cdot 2a_3 = a_1$
x^2	$4 \cdot 3a_4 = 2a_2$
x^3	$5 \cdot 4a_5 = 3a_3$

In general, $n(n-1)a_n = (n-2)a_{n-2}$.

$a_n = \frac{(n-2)a_{n-2}}{n(n-1)}$

Thus a_0 is arbitrary, a_1 is arbitrary, $a_2 = 0$,
$a_3 = \frac{a_1}{3\cdot2}$, $a_4 = 0$,

$a_5 = \frac{3a_3}{5\cdot4} = \frac{3a_1}{5\cdot4\cdot3\cdot2}$, $a_6 = 0$,

$a_7 = \frac{5a_5}{7\cdot6} = \frac{5\cdot3a_1}{7\cdot6\cdot5\cdot4\cdot3\cdot2}$, etc.

In general

$a_n = \begin{cases} 0, & n \text{ even} \\ \frac{(n-2)(n-4)\cdots1}{n!}a_1, & n \text{ odd} \end{cases}$

or, equivalently, $a_{2n} = 0$, and

$a_{2n+1} = \frac{(2n-1)\cdots3\cdot1}{n!}a_1$ for all n.

Thus $y = a_0 + a_1 \sum_{n=1}^{\infty} \frac{(2n-1)\cdots3\cdot1}{(2n+1)!}x^{2n+1}$ is the

solution to $y'' = xy'$, where a_0 and a_1 are arbitrary constants.

Section 10.1

_____ **1.** $\lim\limits_{n \to \infty} \frac{n^2+3n}{2n^2+n+1} =$

  a) 0   b) $\frac{1}{2}$ c) 1   d) ∞

_____ **2.** Sometimes, Always, or Never:
If $a_n \geq b_n \geq 0$ and $\{b_n\}$ diverges, then $\{a_n\}$ diverges.

_____ **3.** Sometimes, Always, or Never:
If $a_n \geq b_n \geq c_n$ and both $\{a_n\}$ and $\{c_n\}$ converge, then $\{b_n\}$ converges.

_____ **4.** Sometimes, Always, or Never:
If $\{a_n\}$ is increasing and bounded above, then $\{a_n\}$ converges.

_____ **5.** True, False:
$a_n = \frac{(-1)^n}{n^2}$ is monotonic.

_____ **6.** $\lim\limits_{n \to \infty} \frac{\arctan n}{2} =$

  a) $\frac{\pi}{4}$ b) $\frac{\pi}{2}$ c) π d) does not exist

_____ **7.** The fourth term of $\{a_n\}$ defined by $a_1 = 3$, and $a_{n+1} = \frac{2}{3}a_n$, $n = 1, 2, 3, \ldots$, is:

  a) $\frac{16}{81}$ b) $\frac{16}{27}$ c) $\frac{8}{27}$ d) $\frac{8}{9}$

_____ 1. True, False:

$$\sum_{n=1}^{\infty} a_n \text{ converges if } \lim_{n \to \infty} a_n = 0.$$

_____ 2. True, False:

If $\sum_{n=1}^{\infty} a_n$ converges and $\sum_{n=1}^{\infty} b_n$ converges, then $\sum_{n=1}^{\infty} (a_n - b_n)$ converges.

_____ 3. The harmonic series is:

a) $1 + 2 + 3 + 4 + \dots$

b) $1 + \frac{1}{2} + \frac{1}{3} + \frac{1}{4} + \dots$

c) $1 + \frac{1}{2} + \frac{1}{4} + \frac{1}{8} + \dots$

d) $1 - \frac{1}{2} + \frac{1}{4} - \frac{1}{8} + \dots$

_____ 4. True, False:

If $\sum_{n=1}^{\infty} \frac{n}{n+1}$ converges.

_____ 5. $\sum_{n=1}^{\infty} 2\left(\frac{1}{4}\right)^n$ converges to:

a) $\frac{9}{4}$ b) 2   c) $\frac{8}{3}$ d) the series diverges

_____ 6. True, False:

$-3 + 1 - \frac{1}{3} + \frac{1}{9} - \frac{1}{27} + \dots$ is a geometric series.

_____ 1. For what values of p does the series $\sum_{n=1}^{\infty} \frac{1}{(n^2)^p}$ converge?

a) $p > -\frac{1}{2}$ b) $p < -\frac{1}{2}$ c) $p > \frac{1}{2}$ d) $p < \frac{1}{2}$

_____ 2. True, False:
If $f(n) = a_n$ for all n, and $f(x)$ is continuous, decreasing, and $\int_1^{\infty} f(x)dx = M$, then $\sum_{n=1}^{\infty} a_n = M$.

_____ 3. Does $\sum_{n=1}^{\infty} \frac{1}{n^{2/3}}$ converge?

_____ 4. Does $\sum_{n=1}^{\infty} \frac{n+2}{(n^2+4n+1)^2}$ converge?

_____ 5. For $s = \sum_{n=1}^{\infty} \frac{1}{n^3}$, an upper bound estimate for $s - s_6$ (where s_6 is the sixth partial sum) is:

a) $\int_1^{\infty} x^{-3}\ dx$

b) $\int_6^{\infty} x^{-3}\ dx$

c) $\int_7^{\infty} x^{-3}\ dx$

d) the series does not converge

_____ 1. Sometimes, Always, or Never:

If $0 \le a_n \le b_n$ for all n and $\sum\limits_{n=1}^{\infty} a_n$ diverges, then $\sum\limits_{n=1}^{\infty} b_n$ converges.

_____ 2. Does $\sum\limits_{n=1}^{\infty} \frac{n+1}{n^3}$ converge?

_____ 3. Does $\sum\limits_{n=1}^{\infty} \frac{\sqrt{n}+\sqrt[3]{n}}{n^{2/3}+n^{3/2}+1}$ converge?

_____ 4. Does $\sum\limits_{n=1}^{\infty} \frac{\cos^2(2^n)}{2^n}$ converge?

_____ 5. $s = \sum\limits_{n=1}^{\infty} \frac{1}{n \cdot 2^n}$ converges by the Comparison Test, comparing it to $\sum\limits_{n=1}^{\infty} \frac{1}{2^n}$.

Using this information, make an estimate of the difference between s and its third partial sum.

a) $\frac{15}{16}$　　　　b) $\frac{7}{8}$　　　　c) $\frac{31}{32}$　　　　d) 1

_____ 1. Does $\displaystyle\sum_{n=1}^{\infty} \frac{(-1)^{n+1} \ln n}{n^2}$ converge?

_____ 2. Does $\displaystyle\sum_{n=1}^{\infty} \frac{(-1)^n}{\sqrt[4]{n+1}}$ converge?

_____ 3. For what value of n is the nth partial sum within 0.001 of the value of $\displaystyle\sum_{n=1}^{\infty} \frac{(-1)^n}{2^n}$?

   a) $n = 4$ b) $n = 6$ c) $n = 8$ d) $n = 10$

_____ 4. True, False:

The Alternating Series Test may be applied to determine the convergence of $\displaystyle\sum_{n=1}^{\infty} \frac{1+(-1)^n}{2n^2}$.

_____ 1. True, False:

If $\sum\limits_{n=1}^{\infty} a_n$ converges absolutely, then it converges conditionally.

_____ 2. True, False:

If $\lim\limits_{n\to\infty} \left| \frac{a_n}{a_{n+1}} \right| = 3$, then $\sum\limits_{n=1}^{\infty} a_n$ converge absolutely.

_____ 3. True, False:

Every series must do one of these: converge absolutely, converge conditionally, or diverge.

_____ 4. The series $\sum\limits_{n=1}^{\infty} 2^{-n} n!$ _____.

    a) diverges
    b) converges absolutely
    c) converges conditionally
    d) converges, but not absolutely or conditionally

_____ 5. The series $\sum\limits_{n=1}^{\infty} \frac{(-5)^{n+1}}{n^n}$ _____.

    a) diverges
    b) converges absolutely
    c) converges conditionally
    d) converges, but not absolutely or conditionally

Section 10.7

_____ **1.** True, False:

$\sum_{n=2}^{\infty} \frac{1}{(\ln n)^n}$ converges.

_____ **2.** True, False:

$\sum_{n=1}^{\infty} \frac{6}{7n+8}$ converges.

_____ **3.** True, False:

$\sum_{n=1}^{\infty} \frac{(-1)^n \sqrt{n}}{n+3}$ converges.

_____ **4.** True, False:

$\sum_{n=1}^{\infty} \frac{e^n}{n!}$ converges.

_____ 1. Sometimes, Always, Never:

The interval of convergence of a power series $\sum_{n=0}^{\infty} a_n(x - c)^n$ is an open interval $(c - R, c + R)$. (When $R = 0$ we mean $\{0\}$ and when $R = \infty$ we mean $(-\infty, \infty)$.)

_____ 2. True, False:

If a number p is in the interval of convergence of $\sum_{n=0}^{\infty} a_n x^n$ then so is the number $\frac{p}{2}$.

_____ 3. For $f(x) = \sum_{n=0}^{\infty} \frac{(x-1)^n}{3^n}$, $f(3) =$

a) 0   b) 2   c) 3   d) $f(3)$ does not exist

_____ 4. The interval of convergence of $\sum_{n=1}^{\infty} \frac{x^n}{\sqrt{n}}$ is:

a) $[-1, 1]$ b) $[-1, 1)$ c) $(-1, 1]$ d) $(-1, 1)$

_____ 5. The radius of convergence of $\sum_{n=0}^{\infty} \frac{n(x-5)^n}{3^n}$ is:

a) $\frac{1}{3}$ b) 1   c) 3   d) ∞

Section 10.9

_____ 1. Given that $e^x = \sum\limits_{n=0}^{\infty} \frac{x^n}{n!}$, a power series for xe^{x^2} is:

a) $\sum\limits_{n=0}^{\infty} \frac{x^2}{n!}$

b) $\sum\limits_{n=0}^{\infty} \frac{x^{2n}}{n!}$

c) $\sum\limits_{n=0}^{\infty} \frac{x^{2n+1}}{n!}$

d) $\sum\limits_{n=0}^{\infty} \frac{x^{2n}}{(n+1)!}$

_____ 2. For $f(x) = \sum\limits_{n=0}^{\infty} \frac{x^{2n}}{n!}$, $f'(x) =$

a) $\sum\limits_{n=1}^{\infty} x^{2n-1}$

b) $\sum\limits_{n=1}^{\infty} 2x^{2n-1}$

c) $\sum\limits_{n=1}^{\infty} 2^n x^{2n-1}$ d)

$\sum\limits_{n=1}^{\infty} (2n-1)x^{2n-1}$

_____ 3. Using $\frac{1}{1-x} = \sum\limits_{n=0}^{\infty} x^n$, $\int \frac{x}{1-x^2} \, dx =$

a) $\sum\limits_{n=0}^{\infty} \frac{x^{2n}}{2n}$

b) $\sum\limits_{n=0}^{\infty} \frac{x^{2n+1}}{2n+1}$

c) $\sum\limits_{n=0}^{\infty} \frac{x^{2n+2}}{2n+2}$

d) $\sum\limits_{n=0}^{\infty} x^{2n+1}$

Section 10.10

_____ 1. Given the Taylor series $e^x = \sum\limits_{n=0}^{\infty} \frac{x^n}{n!}$, a Taylor series for $e^{x/2}$ is:

a) $\sum\limits_{n=0}^{\infty} \frac{2^n x^n}{n!}$

b) $\sum\limits_{n=0}^{\infty} \frac{2x^n}{n!}$

c) $\sum\limits_{n=0}^{\infty} \frac{x^n}{2^n n!}$

d) $\sum\limits_{n=0}^{\infty} \frac{x^n}{2 n!}$

_____ 2. The nth term in the Taylor series about $x = 1$ for $f(x) = x^{-2}$ is:

a) $(-1)^{n+1}(n+1)!(x-1)^n$

b) $(-1)^n n!(x-1)^n$

c) $(-1)^{n+1}(n+1)(x-1)^n$

d) $(-1)^n n(x-1)^n$

_____ 3. The Taylor polynomial of degree 3 for $f(x) = x(\ln x - 1)$ about $x = 1$ is $T_3(x) =$

a) $-1 + \frac{x^2}{x} - \frac{x^3}{6}$

b) $-1 + x - \frac{x^2}{2} + \frac{x^3}{6}$

c) $-1 + x - x^2 + x^3$

d) $-1 - x + x^2 - x^3$

_____ 4. True, False:
If $T_n(x)$ is the nth Taylor polynomial for $f(x)$ at $x = c$, then $T_n^{(k)}(c) = f^{(k)}(c)$ for $k = 0, 1, \ldots, n$.

_____ 5. Taylor's formula with $n = 3$ for $f(x) = \sqrt{x}$ about $c = 9$ is:

a) $-\frac{5(x-9)^3}{128z^{5/2}}$

b) $-\frac{5(x-9)^4}{128z^{7/2}}$

c) $-\frac{15(x-9)^3}{16z^{5/2}}$

d) $-\frac{15(x-9)^4}{16z^{7/2}}$

Section 10.11

_____ 1. $\left(\begin{smallmatrix} \frac{1}{2} \\ 3 \end{smallmatrix}\right) =$

     a) $\frac{5}{16}$ b) $-\frac{5}{16}$ c) $\frac{5}{8}$ d) $-\frac{5}{8}$

_____ 2. $\displaystyle\sum_{n=0}^{\infty} \left(\begin{smallmatrix} \frac{2}{3} \\ n \end{smallmatrix}\right) x^n$ is the binomial series for:

     a) $(1+x)^{2/3}$ b) $(1+x)^{-2/3}$

     c) $(1+x)^{3/2}$ d) $(1+x)^{-3/2}$

_____ 3. Using a binomial series, the Maclaurin series for $\frac{1}{1+x^2}$ is:

     a) $\displaystyle\sum_{n=0}^{\infty} \left(\begin{smallmatrix} 1 \\ n \end{smallmatrix}\right) x^n$ b) $\displaystyle\sum_{n=0}^{\infty} \left(\begin{smallmatrix} -1 \\ n \end{smallmatrix}\right) x^n$

     c) $\displaystyle\sum_{n=0}^{\infty} \left(\begin{smallmatrix} 1 \\ n \end{smallmatrix}\right) x^{2n}$ d) $\displaystyle\sum_{n=0}^{\infty} \left(\begin{smallmatrix} -1 \\ n \end{smallmatrix}\right) x^{2n}$

Section 10.12

_____ **1.** If the Maclaurin polynomial of degree 2 for $f(x) = e^x$ is used to approximate $e^{0.2}$ then an estimate for the error with $0 < z < 0.2$ is:

a) $\frac{1}{24}e^z$ b) $\frac{1}{6}e^z$ c) $\frac{1}{2}e^z$ d) e^z

_____ **2.** Let $f(x) = e^{2x}$, $c = 0$. Use Taylor's formula to estimate the accuracy of $f(x) \approx T_4(x)$ when $-1 \le x \le 1$.

a) $\frac{4e^2}{15}$ b) $\frac{4}{15}$ c) $\frac{e^2}{3}$ d) $\frac{1}{3}$

_____ 1. The point plotted at the right
has coordinates:

a) (2, 5, 4)   b) (5, 2, 4)
c) (4, 2, 5)   d) (2, 4, 5)

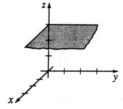

_____ 2. The equation of the plane
partially drawn is:

a) $x = 3$ b) $y = 3$
c) $z = 3$ d) $x + y = 3$

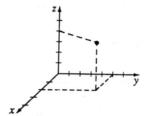

_____ 3. The distance between $(7, 4, -3)$ and $(-1, 2, 3)$ is:

a) 4   b) 16   c) 104   d) $\sqrt{104}$

_____ 4. The sphere graphed at the
right has equation

a) $(x - 2)^2 + (y - 3)^2 + (z - 4)^2 = 1$
b) $(x - 2)^2 + (y - 4)^2 + (z - 4)^2 = 4$
c) $(x - 2)^2 + (y - 5)^2 + (z - 4)^2 = 4$
d) $(x - 2)^2 + (y - 4)^2 + (z - 4)^2 = 1$

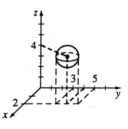

_____ 5. The region described by $y^2 + z^2 = 4$ is:

a) a circle, center at the origin   b) a sphere, center at the origin
c) a cone, along the x-axis   d) a cylinder, along the x-axis

Section 11.2

_____ **1.** The vector represented by $\overrightarrow{AB}$ where A: $(4, 8)$, B: $(6, 6)$ is:

 a) $\langle -2, 2 \rangle$ b) $\langle 2, -2 \rangle$ c) $\langle 10, 14 \rangle$ d) $\langle 14, 10 \rangle$

_____ **2.** The length of $\mathbf{a} = 4\mathbf{i} - \mathbf{j} - 2\mathbf{k}$ is:

 a) $\sqrt{11}$ b) 11 c) $\sqrt{21}$ d) 21

_____ **3.** For $\mathbf{a} = 6\mathbf{i} - \mathbf{j}$ and $\mathbf{b} = 2\mathbf{i} + 3\mathbf{j}$, $\mathbf{a} + 2\mathbf{b} =$

 a) $2\mathbf{i} - 4\mathbf{j}$ b) $8\mathbf{i} - 2\mathbf{j}$ c) $16\mathbf{i} - 4\mathbf{j}$ d) $10\mathbf{i} + 5\mathbf{j}$

_____ **4.** True, False:
 $\mathbf{i} + \mathbf{j}$ is a unit vector in the direction of $5\mathbf{i} + 5\mathbf{j}$.

_____ **5.** The vector **c** in the figure at
 the right is:

 a) $\mathbf{a} - \mathbf{b}$
 b) $\mathbf{b} - \mathbf{a}$
 c) $\mathbf{a} + \mathbf{b}$
 d) $-\mathbf{a} - \mathbf{b}$

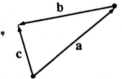

_____ **1.** For $\mathbf{a} = 2\mathbf{i} + 3\mathbf{j} - 4\mathbf{k}$ and $\mathbf{b} = -\mathbf{i} + 2\mathbf{j} - \mathbf{k}$, $\mathbf{a} \cdot \mathbf{b} =$

     a) 0   b) 8   c) 10   d) 12

_____ **2.** The angle θ at the right has cosine:

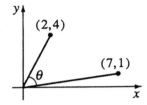

     a) $\dfrac{9}{5\sqrt{5}}$ b) $\dfrac{9}{250}$

     c) $\dfrac{9}{\sqrt{50}}$ d) $\dfrac{9}{50}$

_____ **3.** The scalar projection of $\mathbf{a} = \langle 1, 4 \rangle$ on $\mathbf{b} = \langle 6, 3 \rangle$ is:

     a) $\dfrac{18}{\sqrt{17}}$ b) $\dfrac{6}{\sqrt{5}}$ c) $\dfrac{6}{\sqrt{17}\sqrt{5}}$ d) $\dfrac{18}{\sqrt{17}\sqrt{5}}$

_____ **4.** The α, β, γ direction cosines in the figure are (respectively):

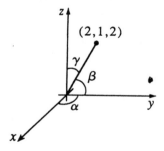

     a) $\dfrac{2}{9}, \dfrac{1}{9}, \dfrac{2}{9}$ b) $\dfrac{2}{3}, \dfrac{1}{3}, \dfrac{2}{3}$

     c) $\sqrt{\dfrac{2}{9}}, \sqrt{\dfrac{1}{9}}, \sqrt{\dfrac{2}{9}}$ d) $\sqrt{\dfrac{2}{3}}, \sqrt{\dfrac{1}{3}}, \sqrt{\dfrac{2}{3}}$

_____ 1. For $\mathbf{a} = \langle 1, 2, 1 \rangle$ and $\mathbf{b} = \langle 3, 0, 2 \rangle$, $\mathbf{a} \times \mathbf{b} =$.

     a) $\langle -6, -1, 4 \rangle$ b) $\langle -6, 1, 4 \rangle$
     c) $\langle 4, -1, -6 \rangle$ d) $\langle 4, 1, -6 \rangle$

_____ 2. True, False:
     $\mathbf{a} \times \mathbf{b}$ is in the plane formed by vectors $\mathbf{a}$ and $\mathbf{b}$.

_____ 3. $\mathbf{j} \times (-\mathbf{k}) =$

     a) $\mathbf{i}$ b) $-\mathbf{i}$ c) $\mathbf{j} + \mathbf{k}$ d) $-\mathbf{j} - \mathbf{k}$

_____ 4. The area of the parallelogram at the right is:

     a) 26
     b) $\sqrt{14}$
     c) $\sqrt{86}$
     d) $\sqrt{300}$

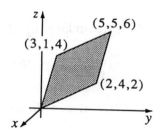

_____ 5. For $\mathbf{a} = \langle 6, 2, 3 \rangle$, $\mathbf{b} = \langle 4, 7, 9 \rangle$, $\mathbf{c} = \langle 8, 1, 5 \rangle$, $\mathbf{c} \cdot (\mathbf{a} \times \mathbf{b}) =$

   a) $\begin{vmatrix} 6 & 2 & 3 \\ 4 & 7 & 9 \\ 8 & 1 & 5 \end{vmatrix}$ b) $\begin{vmatrix} 6 & 2 & 3 \\ 8 & 1 & 5 \\ 4 & 7 & 9 \end{vmatrix}$ c) $\begin{vmatrix} 8 & 1 & 5 \\ 6 & 2 & 3 \\ 4 & 7 & 9 \end{vmatrix}$ d) $\begin{vmatrix} 8 & 1 & 5 \\ 4 & 7 & 9 \\ 6 & 2 & 3 \end{vmatrix}$

Section 11.5

_____ **1.** The equation of the line through $(4, 10, 8)$ and $(3, 5, 1)$ is:

     a) $\frac{x-4}{3} = \frac{y-10}{5} = \frac{z-8}{1}$ b) $\frac{x-3}{4} = \frac{y-5}{10} = \frac{z-1}{8}$

     c) $\frac{x-3}{7} = \frac{y-5}{15} = \frac{z-1}{9}$ d) $\frac{x-4}{1} = \frac{y-10}{5} = \frac{z-8}{7}$

_____ **2.** True, False:

These lines are skew: $x = 1 + 5t$ $x = 4 + 7s$
                              $y = 3 - 6t$ $y = 5 + s$
                              $z = 1 - 2t$ $z = 1 - 4s$

_____ **3.** The equation of the plane through $\langle -7, 2, 3 \rangle$ with normal vector $\langle 6, -4, 1 \rangle$ is:

     a) $6x - 4y + z = -47$ b) $6x - 4y + z = 0$
     c) $-7x + 2y + 3 = -20$ d) $-7x + 2y + 3 = -12$

_____ **4.** The equation of the plane formed by the two lines

$x = 3 + 2t$ $x = 3 + t$
$y = 1 - 4t$ and $y = 1 + 2t$
$z = 5 + t$ $z = 5 + 2t$
is

     a) $2(x - 3) - 4(y - 1) + (z - 5) = 0$
     b) $(x - 3) + 2(y - 1) + 2(z - 5) = 0$
     c) $(x - 3) - 6(y - 1) + (z - 5) = 0$
     d) $-10(x - 3) - 3(y - 1) + 8(z - 5) = 0$

_____ **5.** The distance from $(1, 2, 1)$ to the plane $6x + 5y + 8z = 34$ is:

     a) $\sqrt{5}$ b) $2\sqrt{5}$ c) $\frac{2}{\sqrt{5}}$ d) $\frac{1}{\sqrt{5}}$

Section 11.6

_____ 1. $\frac{x^2}{25} + 1 = \frac{z^2}{16} - \frac{y^2}{9}$ is a(n):

     a) hyperboloid of one sheet   b) hyperboloid of two sheets
     c) ellipsoid   d) hyperbolic cone

_____ 2. $\frac{x^2}{9} + \frac{z^2}{4} = 1$ is a(n):

     a) ellipsoid   b) elliptic cone
     c) elliptic paraboloid   d) elliptic cylinder

_____ 3. The graph at the right has equation

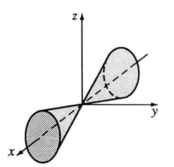

     a) $\frac{x^2}{a^2} + \frac{y^2}{b^2} = \frac{z^2}{c^2}$
     b) $\frac{x^2}{a^2} + \frac{z^2}{c^2} = \frac{y^2}{b^2}$
     c) $\frac{y^2}{b^2} + \frac{z^2}{c^2} = \frac{x^2}{a^2}$
     d) $\frac{x^2}{a^2} + \frac{y^2}{b^2} + \frac{z^2}{c^2} = 0$

_____ 4. The graph at the right has equation

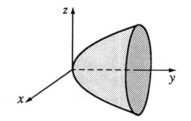

     a) $x = \frac{y^2}{b^2} + \frac{z^2}{c^2}$
     b) $y = \frac{x^2}{a^2} + \frac{z^2}{c^2}$
     c) $z = \frac{x^2}{a^2} + \frac{y^2}{b^2}$
     d) $y^2 = \frac{x^2}{a^2} + \frac{z^2}{c^2}$

_____ 5. $x^2 + 6x + y^2 - 10y + z = 0$ is a(n):

     a) ellipsoid   b) paraboloid
     c) hyperbolic cone   d) elliptic cone

Section 11.7

_____ **1.** The curve given by $\mathbf{r}(t) = 2\mathbf{i} + t\mathbf{j} + 2t\mathbf{k}$ is a

    a) line   b) plane   c) spiral   d) circle

_____ **2.** $\lim\limits_{t \to \infty} \left\langle e^{-2t},\ \cos \frac{1}{t} \right\rangle =$

    a) $\langle 1,\ 1 \rangle$ b) $\langle 0,\ 1 \rangle$ c) $\langle 0,\ 0 \rangle$ d) does not exist

_____ **3.** For $\mathbf{r}(t) = t^3\mathbf{i} + \sin t\mathbf{j} - (t^2 + 2t)\mathbf{k}$, $\mathbf{r}'(0) =$

    a) $\mathbf{j} - 2\mathbf{k}$ b) $3\mathbf{i} - 4\mathbf{k}$ c) $3\mathbf{i} + \mathbf{j} - 2\mathbf{k}$ d) $\mathbf{0}$ (zero vector)

_____ **4.** True, False:
$[\mathbf{r}(t) \times \mathbf{s}(t)]' = \mathbf{r}(t) \times \mathbf{s}'(t) + \mathbf{s}(t) \times \mathbf{r}'(t)$.

_____ **5.** $\int_0^1 (e^t\mathbf{i} + 3\sqrt{t}\mathbf{j} + 2t\mathbf{k})dt =$

    a) $e\mathbf{i} + 2\mathbf{j} + \mathbf{k}$
    b) $e\mathbf{i} + \frac{3}{2}\mathbf{j} + 2\mathbf{k}$
    c) $(e - 1)\mathbf{i} + 2\mathbf{j} + \mathbf{k}$
    d) $(e - 1)\mathbf{i} + \frac{3}{2}\mathbf{j} + 2\mathbf{k}$

_____ **1.** A definite integral for the length of the curve given by
$\mathbf{r}(t) = \langle t,\ 3+t,\ t^2 \rangle$ for $1 \le t \le 2$ is:

    a) $\int_1^2 \sqrt{2 + 4t^2}\ dt$ b) $\int_1^2 \sqrt{9 + 6t + 2t^2 + t^4}\ dt$

    c) $\int_1^2 \sqrt{9 + 4t}\ dt$ d) $\int_1^2 2t\ dt$

_____ **2.** For the vector function $\mathbf{r}(t)$ with unit tangent $\mathbf{T}$, the unit normal is:

    a) $\frac{\mathbf{T}}{|\mathbf{T}|}$ b) $\frac{\mathbf{r}'}{|\mathbf{r}'|}$ c) $\frac{\mathbf{r}''}{|\mathbf{r}''|}$ d) $\frac{\mathbf{T}'}{|\mathbf{T}'|}$

_____ **3.** The curvature of $\mathbf{r}(t) = -e^t\mathbf{i} + t\mathbf{j} + e^t\mathbf{k}$ at $t = 0$ is:

    a) $\frac{\sqrt{2}}{3\sqrt{3}}$ b) $\frac{\sqrt{2e}}{(2e+1)^{3/2}}$ c) $\frac{1}{(2e)^{3/2}}$ d) $\frac{\sqrt{2e}}{(2e^2+1)^{3/2}}$

_____ **4.** True, False:
If f is twice differentiable and x_0 is an inflection point for f then the
curvature of f at x_0 is 0.

_____ **5.** The normal plane to $\mathbf{r}(t) = t^2\mathbf{i} - t^3\mathbf{j} + t^4\mathbf{k}$ at $t = 1$ has equation:

    a) $2(x - 1) - 3(y + 1) + 4(z - 1) = 9$
    b) $2(x - 1) - 3(y + 1) + 4(z - 1) = 20$
    c) $(x - 2) - (y + 3) + (z - 4) = 9$
    d) $(x - 2) - (y + 3) + (z - 4) = 20$

_____ 1. The acceleration for a particle whose position at time t is
$3t^3\mathbf{i} + \ln t\mathbf{j} - \sin 2t\mathbf{k}$ is:

     a) $9t^2\mathbf{i} + \frac{1}{t}\mathbf{j} - 2\cos 2t\mathbf{k}$ b) $18t\mathbf{i} - \frac{1}{t^2}\mathbf{j} + 4\sin 2t\mathbf{k}$

     c) $18\mathbf{i} + \frac{2}{t^3}\mathbf{j} - 8\cos 2t\mathbf{k}$ d) $\frac{3}{4}t^4\mathbf{i} + (t\ln t - t)\mathbf{j} + \frac{1}{2}\cos 2t\mathbf{k}$

_____ 2. With initial position $\mathbf{i}$, velocity $2\mathbf{j}$, and acceleration $30t\mathbf{i}+60t^2\mathbf{k}$, the position function is:

     a) $5t^3\mathbf{i} + 2\mathbf{j} + 5t^4\mathbf{k}$ b) $5t^3\mathbf{i} + (2t + 1)\mathbf{j} + 5t^4\mathbf{k}$

     c) $(5t^3 + 1)\mathbf{i} + 2t\mathbf{j} + 5t^4\mathbf{k}$ d) $5t^3\mathbf{i} + 2\mathbf{j} + (5t^4 + 1)\mathbf{k}$

_____ 3. The force needed for a 10 kg object to attain velocity $6t^3\mathbf{i} + 10t\mathbf{j}$ is:

     a) $20t^3\mathbf{i} + 50t^2\mathbf{j}$ b) $5t^4\mathbf{i} + \frac{50}{3}t^3\mathbf{j}$

     c) $180t^2\mathbf{i} + 100\mathbf{j}$ d) $120\mathbf{i}$

_____ 4. The tangential component of acceleration for $\mathbf{r}(t) = e^t\mathbf{i} - e^{-t}\mathbf{j}$ at $t = 0$ is:

     a) 0 b) $\frac{e^4 - 1}{\sqrt{e^6 - e^2}}$ c) $\frac{e^4}{\sqrt{e^2 - 1}}$ d) $\frac{e^4 - 1}{e}$

_____ **1.** The spherical coordinates of the point with rectangular coordinates $(-2,\ 2\sqrt{3},\ 4\sqrt{3})$ are:

    a) $\left(8,\ \frac{2\pi}{3},\ \frac{\pi}{6}\right)$ b) $\left(8,\ \frac{\pi}{3},\ -\frac{\pi}{6}\right)$

    c) $\left(8,\ \frac{\pi}{3},\ \frac{\pi}{6}\right)$ d) $\left(8,\ \frac{2\pi}{3},\ -\frac{\pi}{6}\right)$

_____ **2.** The rectangular coordinates of the point with spherical coordinates $\left(2,\ \frac{\pi}{2},\ \frac{\pi}{4}\right)$ are:

    a) $(2,\ \sqrt{2},\ \sqrt{2})$ b) $(\sqrt{2},\ \sqrt{2},\ 0)$

    c) $(\sqrt{2},\ 0,\ \sqrt{2})$ d) $(0,\ \sqrt{2},\ \sqrt{2})$

_____ **3.** The surface given by $z = -r^2$ is a(n)

    a) elliptic cone   b) hyperboloid of one sheet

    c) hyperbolic cylinder   d) elliptic paraboloid

_____ **4.** The point P graphed at the right has spherical coordinates

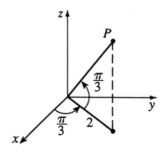

    a) $\left(2,\ \frac{\pi}{3},\ \frac{\pi}{3}\right)$

    b) $\left(2,\ \frac{\pi}{3},\ \frac{\pi}{6}\right)$

    c) $\left(4,\ \frac{\pi}{3},\ \frac{\pi}{3}\right)$

    d) $\left(4,\ \frac{\pi}{3},\ \frac{\pi}{6}\right)$

_____ **5.** The cylindrical coordinates of the point with rectangular coordinates $(5,\ 5,\ 4)$ are:

    a) $\left(5\sqrt{2},\ \frac{\pi}{4},\ 4\right)$ b) $\left(5\sqrt{2},\ \frac{\pi}{2},\ 4\right)$ c) $\left(5,\ \frac{\pi}{4},\ 4\right)$ d) $\left(5,\ \frac{\pi}{2},\ 4\right)$

Section 12.1

_____ 1. Which of these points is not in the domain of $f(x, y) = \sqrt{8 - x - 2y^2}$?

    a) $(0, 0)$ b) $(-6, 2)$

    ⁀ c) $(3, 3)$ d) All the points _are_ in the domain.

_____ 2. The level curves (for $k \neq 0$) of $f(x, y) = xy$ are:

    a) ellipses   b) hyperbolas

    c) parabolas   d) pairs of lines

_____ 3. Which function best fits the graph at the right?

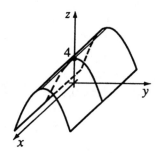

    a) $f(x, y) = 4 - x^2$
    b) $f(x, y) = 4 - y^2$
    c) $f(x, y) = 4 - xy$
    d) $f(x, y) = 4 - x^2 - y^2$

_____ 4. The range of $f(x, y) = \sqrt{x} + \frac{1}{\sqrt{y}}$ is:

    a) $(-\infty, \infty)$ b) $[0, \infty)$

    c) $(0, \infty)$ d) $[1, \infty]$

_____ **1.** True, False:
If $\lim\limits_{(x,y)\to(a,b)} f(x,y) = L$ and $\lim\limits_{(x,y)\to(a,b)} g(x,y) = M \neq 0$, then
$\lim\limits_{(x,y)\to(a,b)} \frac{f(x,y)}{g(x,y)} = \frac{L}{M}$.

_____ **2.** $\lim\limits_{(x,y)\to(-1,2)} \frac{xy}{x^2-y^2} =$

    a) $\frac{2}{3}$ b) $-\frac{2}{3}$ c) $-\frac{2}{5}$ d) does not exist

_____ **3.** $\lim\limits_{(x,y)\to(0,0)} \frac{\sin(xy)}{y^2} =$

    a) 0   b) 1   c) π d) does not exist

_____ **4.** $f(x,y) = \sqrt{x+y}$ is continuous for all (x,y) such that:

    a) $x \geq 0$ and $y \geq 0$ b) $y > x$
    c) $y > -x$ d) $-y \leq x$

_____ **5.** True, False:
$f(x,y) = \begin{cases} \frac{\sin x}{y} & y \neq 0 \\ 1 & y = 0 \end{cases}$ is continuous at $(0,0)$.

Section 12.3

_____ 1. $f_x(a, b)$ is the slope of the tangent line to the surface $z = f(x, y)$ at (a, b) determined by:

   a) the trace through $x = a$
   b) the trace through $y = b$
   c) the trace through $x = y$
   d) the intersection of the two traces $x = a$ and $y = b$

_____ 2. For $f(x, y) = e^{x^2 y^3}$, $f_y(x, y) =$

   a) $3y^2 e^{x^2 y^3}$

   b) $3x^2 y^2 e^{x^2 y^3}$

   c) $(2xy^3 + 3x^2 y^3) e^{x^2 y^3}$

   d) $e^{3x^2 y^3}$

_____ 3. For $f(x, y) = \ln(2x + 3y)$, $f_{yx} =$

   a) $\frac{-6}{(2x+3y)^2}$

   b) $\frac{3}{2x+3y}$

   c) $\frac{6}{2x+3y}$

   d) $\frac{2}{x} + \frac{3}{y}$

_____ 4. Sometimes, Always, or Never:
   $f_{xy} = f_{yz}$.

_____ 5. How many second partial derivatives will there be for $z = f(r, s, t, x, y)$? (Do not assume they are continuous.)

   a) 5   b) 10   c) 20   d) 25

_____ **1.** True, False:
If dz exists at (x, y) then $z = f(x, y)$ is differentiable at (x, y).

_____ **2.** The equation of the tangent plane to $z = x^2 + xy + y^3$ for $(x, y) = (2, -1)$ is:

a) $3x + 5y - z = 10$ b) $3x + 5y - z = 2$
c) $5x + 3y - z = 10$ d) $5x + 3y - z = 2$

_____ **3.** For $z = \sin(xy)$, $dz =$

a) $xy\cos(xy)dx + xy\cos(xy)dy$
b) $x\cos(xy)dx + y\cos(xy)dy$
c) $y\cos(xy)dx + x\cos(xy)dy$
d) 0

_____ **4.** The volume of a pyramid is $\frac{1}{3}$ area of its base times its altitude. Using differentials, an estimate for the volume of the pyramid at the right is:

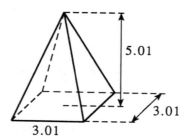

5.01

3.01

3.01

a) 15.10
b) 15.12
c) 15.13
d) 15.14

Section 12.5

_____ **1.** For $z = xy^2$, $x = t + v^2$, $y = t^2 + v$, $\frac{\partial z}{\partial v} =$

    a) $2[(t^2 + v)^2 + (t^2 + v)(t + v^2)]$
    b) $2[(t^2 + v)^2 + (t + v^2)^2]$
    c) $2(t^2 + v)[2t^2v + v^2]$
    d) $2(t^2 + v)[t^2v + t + 2v^2]$

_____ **2.** Find $\frac{\partial f}{\partial r}$ for $f(x, y) = x^2 + 2y^2$, $x = 2rs$, $y = 4r^2s^2$.

    a) $8rs^2 + 128r^3s^4$ b) $2r^2s^2 + 64r^3s^4$
    c) $4rs^2 + 64r^3s^4$ d) $2r^2s^2 + 128r^3s^4$

_____ **3.** Find $\frac{dy}{dx}$ for y defined by $x^2 - 6xy + y^2 = 20$.

    a) $\frac{x-3y}{3x-y}$ b) $\frac{x-3y}{y-3x}$ c) $\frac{y-3x}{x-3y}$ d) $\frac{3x-y}{x-3y}$

_____ **4.** Find $\frac{\partial z}{\partial x}$ for $z = f(x, y)$ defined implicitly by $\frac{x^2}{4} - \frac{y^2}{9} + \frac{z^2}{16} = 0$.

    a) $\frac{-x}{4z}$ b) $\frac{x}{4z}$ c) $\frac{-4x}{z}$ d) $\frac{4x}{z}$

_____ **1.** For $f(x, y) = x^3y + xy^2$ and $\mathbf{u} = \left\langle \frac{3}{5}, -\frac{4}{5} \right\rangle$, $D_{\mathbf{u}}f(-1, 1) =$

     a) $\frac{12}{5}$ b) $-\frac{12}{5}$ c) $\frac{24}{5}$ d) $-\frac{24}{5}$

_____ **2.** True, False:
$D_{\mathbf{u}}f(a, b) = \nabla f(a, b) \cdot \mathbf{u}$.

_____ **3.** In what direction $\mathbf{u}$ is $D_{\mathbf{u}}f(-1, 1)$ maximum for $f(x, y) = x^3y^4$?

     a) $\langle 3, -4 \rangle$ b) $\left\langle \frac{3}{5}, -\frac{4}{5} \right\rangle$ c) $\langle 4, -3 \rangle$ d) $\left\langle \frac{4}{5}, -\frac{3}{5} \right\rangle$

_____ **4.** The equation of the plane tangent to $x^2 + 2y^2 + 2z^2 = 5$ at $(-1, -1, 1)$ is:

     a) $-(x + 1) - (y + 1) + (z - 1) = 0$
     b) $-(x - 1) - (y - 1) + (z + 1) = 0$
     c) $-2(x + 1) - 4(y + 1) + 4(z - 1) = 0$
     d) $-4(x + 1) - 4(y + 1) - 2(z - 1) = 0$

_____ **5.** Which vector at the right could be a gradient for $f(x, y)$?

     a)   **a**
     b)   **b**
     c)   **c**
     d)   **d**

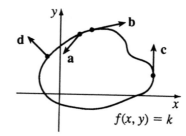

$f(x, y) = k$

_____ 1. A critical point of $f(x, y) = 60 - 6x + x^2 + 8y - y^2$ is:

    a) $(3, 4)$ b) $(-3, -4)$ c) $(-3, 4)$ d) $(3, -4)$

_____ 2. If (a, b) is a critical point of $f(x, y)$, all second derivatives of f are continuous, and $f_{xx}(a, b) = 10$, $f_{yy}(a, b) = 3$, $f_{xy}(a, b) = 5$, then (a, b) is a:

    a) local maximum
    b) local minimum
    c) saddle point
    d) cannot tell from the information given

_____ 3. The absolute minimum of $f(x, y) = 40 - 3x^2 - 2y^2$ with domain $D = \{(x, y)|-2 \le x \le 3, -1 \le y \le 2\}$ is:

    a) 0   b) 5   c) 12   d) 40

_____ 4. A box with a half a lid is to hold 48 cubic inches. If the box is to have minimum surface area, then the value of y must be:

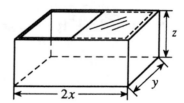

    a) 4   b) $\sqrt[3]{48}$
    c) 6   d) 8

_____ 5. True, False:
    If $f_{xx}f_{yy} - (f_{xy})^2 = 0$ at (x_0, y_0) then (x_0, y_0) cannot be a saddle point.

_____ 1. The absolute maximum of $f(x, y) = 12x - 8y$ subject to $x^2 + 2y^2 = 11$ occurs at:

    a) $(3, -1)$ b) $(-3, 1)$ c) $(\sqrt{11}, 0)$ d) $\left(0, -\sqrt{\frac{11}{2}}\right)$

_____ 2. To find the minimum surface area of a rectangular box whose volume is 100 cm^3 and total edge length is 50 cm by Lagrange multipliers, which set of equations results?

    a) $2y + 2z = \lambda_1 yz + 4\lambda_2$
           $2x + 2z = \lambda_1 xz + 4\lambda_2$
           $2x + 2y = \lambda_1 xy + 4\lambda_2$
           $xyz = 100$
           $4x + 4y + 4z = 50$

    b) $yz = \lambda_1(2y + 2z) + 4\lambda_2$
           $xz = \lambda_1(2x + 2z) + 4\lambda_2$
           $xy = \lambda_1(2x + 2y) + 4\lambda_2$
           $xyz = 100$
           $4x + 4y + 4z = 50$

    c) $4y + 4z = \lambda_1 yz + \lambda_2(2y + 2z)$
           $4x + 4z = \lambda_1 xz + \lambda_2(2x + 2z)$
           $4x + 4y = \lambda_1 xy + \lambda_2(2x + 2z)$
           $xyz = 100$
           $4x + 4y + 4z = 50$

    d) $4 = \lambda_1 xy + \lambda_2(2y + 2z)$
           $4 = \lambda_1 xz + \lambda_2(2x + 2z)$
           $4 = \lambda_1 xy + \lambda_2(2x + 2y)$
           $xyz = 100$
           $4x + 4y + 4z = 50$

Section 13.1

_____ 1. How many subrectangles of $[1, 4] \times [3, 8]$ are determined by the partitions $x_0 = 1$, $x_1 = 2$, $x_2 = 4$ and $y_0 = 3$, $y_1 = 4$, $y_2 = 6$, $y_3 = 8$?

    a) 6   b) 9   c) 12   d) 15

_____ 2. Find the Riemann sum for $\iint_R (x + y)dA$ using midpoints and the partition of R given at the right.

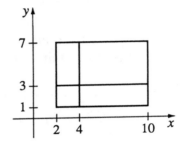

    a) 220
    b) 268
    c) 480
    d) 500

_____ 3. Sometimes, Always, or Never:
$\iint_R f(x, y)dA$ is the volume above the rectangular region R and under $z = f(x, y)$.

_____ 1. Evaluate $\int_0^2 \int_0^1 6xy^2\ dx\ dy$.

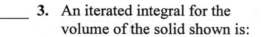

     a) 6   b) 8   c) 24   d) 30

_____ 2. True, False:

$\int_1^5 \int_2^4 (x^2 + y^2)dx\ dy = \int_2^4 \int_1^5 (x^2 + y^2)dx\ dy.$

_____ 3. An iterated integral for the volume of the solid shown is:

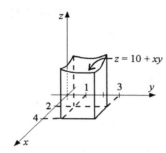

$z = 10 + xy$

     a) $\int_2^4 \int_1^3 (10 + xy)dx\ dy$

     b) $\int_2^4 \int_1^3 (10 + xy)dy\ dx$

     c) $\int_1^2 \int_3^4 (10 + xy)dx\ dy$

     d) $\int_1^2 \int_3^4 (10 + xy)dy\ dx$

_____ 1. Write $\iint_D x \, dA$ as an iterated integral, where D is shown at the right.

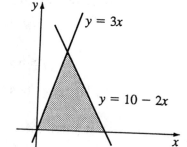

a) $\int_0^2 \int_{10-2x}^{3x} x \, dx \, dy$

b) $\int_0^5 \int_{3x}^{10-2x} x \, dx \, dy$

c) $\int_0^2 \int_{(y+10)/2}^{x/3} x \, dx \, dy$

d) $\int_0^6 \int_{y/3}^{(y+10)/2} x \, dx \, dy$

_____ 2. True, False:
$\int_1^3 \int_2^5 xy^2 \, dy \, dx = \int_2^5 \int_1^3 xy^2 \, dy \, dx.$

_____ 3. Rewritten in reverse order, $\int_0^1 \int_{y-1}^0 x^2 y^2 \, dx \, dy =$

a) $\int_{-1}^1 \int_0^{x-1} x^2 y^2 \, dy \, dx$ b) $\int_{-1}^0 \int_0^{x+1} x^2 y^2 \, dy \, dx$

c) $\int_0^1 \int_0^{x+1} x^2 y^2 \, dy \, dx$ d) $\int_0^1 \int_0^{x-1} x^2 y^2 \, dy \, dx$

_____ 4. The area of the region at the right is:

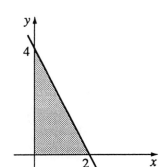

a) $\int_0^2 \int_0^4 dy \, dx$

b) $\int_0^2 \int_0^{4-2x} dy \, dx$

c) $\int_0^4 \int_0^{(y-4)/2} dx \, dy$

d) $\int_0^4 \int_0^2 dx \, dy$

_____ **1.** Rewritten as an iterated integral in polar coordinates

$$\int_0^2 \int_0^{\sqrt{4-x^2}} y \ dy \ dx =$$

a) $\int_0^2 \int_0^{\pi/2} r \sin\theta \ d\theta \ dr$

b) $\int_0^2 \int_0^{\pi/2} r^2 \sin\theta \ d\theta \ dr$

c) $\int_0^{\pi/2} \int_0^2 r^2 \cos\theta \ dr \ d\theta$

d) $\int_0^{\pi/2} \int_0^2 r \cos\theta \ dr \ d\theta$

_____ **2.** An iterated integral for the area of the shaded region D is:

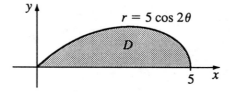

a) $\int_0^5 \int_0^r 5 \cos 2\theta \ d\theta \ dr$

b) $\int_0^5 \int_0^r r \ d\theta \ dr$

c) $\int_0^{\pi/4} \int_0^{5\cos 2\theta} dr \ d\theta$

d) $\int_0^{\pi/4} \int_0^{5\cos 2\theta} r \ dr \ d\theta$

1. The mass of the lamina at the right which has a density of $x^2 + y^2$ at each point (x, y) is given by:

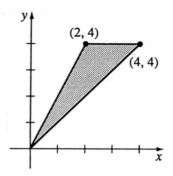

a) $\int_0^4 \int_{y/2}^y (x^2 + y^2) dx \, dy$

b) $\int_0^4 \int_x^{2x} (x^2 + y^2) dy \, dx$

c) $\int_0^4 \int_y^{2y} (x^2 + y^2) dx \, dy$

d) $\int_2^4 \int_{y/2}^y (x^2 + y^2) dx \, dy$

2. The y coordinate of the center of mass of a lamina D with density $\rho(x, y)$ and mass m is:

a) $\iint_D y\rho(x, y) dA$

b) $\iint_D x\rho(x, y) dA$

c) $\dfrac{\iint_D y\rho(x,y) dA}{m}$

d) $\dfrac{\iint_D x\rho(x,y) dA}{m}$

_____ **1.** Write an iterated integral for the area of the plane $z = x$ that is above the region D below.

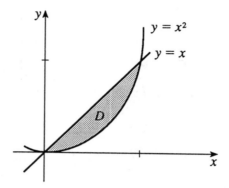

a) $\int_0^1 \int_x^{x^2} \sqrt{x^2 + 1} \, dy \, dx$

b) $\int_0^1 \int_{x^2}^x \sqrt{x^2 + 1} \, dy \, dx$

c) $\int_0^1 \int_{x^2}^x \sqrt{2} \, dy \, dx$

d) $\int_0^1 \int_y^{\sqrt{y}} \sqrt{y + 1} \, dx \, dy$

_____ **1.** True, False:
$\int_3^6 \int_1^5 \int_2^4 xyz \, dx \, dz \, dy = \int_1^4 \int_3^6 \int_2^5 xyz \, dy \, dx \, dz$.

_____ **2.** $\int_1^2 \int_0^x \int_y^{x+y} 12x \, dz \, dy \, dx =$

    a) 45   b) 15   c) 63   d) 127

_____ **3.** $\iiint\limits_E z \, dV$, where E is the
wedge-shaped solid shown
at the right, equals:

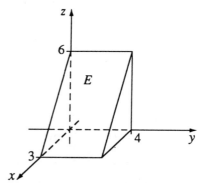

    a) $\int_0^4 \int_0^3 \int_0^{6-2x} z \, dz \, dx \, dy$

    b) $\int_0^4 \int_0^3 \int_0^6 z \, dz \, dx \, dy$

    c) $\int_0^4 \int_0^3 \int_0^{6-z} z \, dx \, dz \, dy$

    d) $\int_0^4 \int_0^3 \int_0^{6-x-y} z \, dz \, dx \, dy$

_____ **4.** The moment about the xz plane of a solid E whose density is
$\rho(x, y, z) = x$ is:

    a) $\iiint\limits_E x^2 z \, dV$ b) $\iiint\limits_E xy \, dV$

    c) $\iiint\limits_E x \, dV$ d) $\iiint\limits_E xyz \, dV$

_____ 1. Write $\iiint\limits_{E} (x^2 + y^2 + z^2)dV$
in cylindrical coordinates,
where E is the solid at the
right.

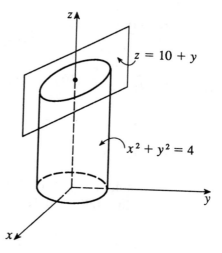

a) $\int_0^{2\pi} \int_0^2 \int_0^{10+y} (r^2 + z^2)r\ dz\ dr\ d\theta$

b) $\int_0^{2\pi} \int_0^4 \int_0^{x^2+y^2} (r^2 + z^2)r\ dz\ dr\ d\theta$

c) $\int_0^{2\pi} \int_0^2 \int_0^{10+y} (r^2 + z^2)\ dz\ dr\ d\theta$

d) $\int_0^{2\pi} \int_0^4 \int_0^{x^2+y^2} (r^2 + z^2)\ dz\ dr\ d\theta$

_____ 2. Write $\iiint\limits_{E} 2\ dV$ in spherical
coordinates, where E is the
bottom half of the sphere at
the right.

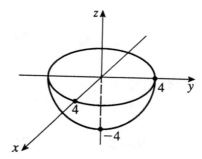

a) $\int_{\pi/2}^{\pi} \int_0^{2\pi} \int_0^2 2\rho^2 \sin\phi\ d\rho\ d\theta\ d\phi$

b) $\int_{\pi/2}^{\pi} \int_0^{2\pi} \int_0^4 2\rho^2 \sin\phi\ d\rho\ d\theta\ d\phi$

c) $\int_{\pi/2}^{\pi} \int_0^{2\pi} \int_0^2 2\ d\rho\ d\theta\ d\phi$

d) $\int_{\pi/2}^{\pi} \int_0^{2\pi} \int_0^4 2\ d\rho\ d\theta\ d\phi$

Section 13.9

_____ 1. Find the Jacobian for $x = u^2v^2$, $y = u^2 + v^2$.

a) $2uv^3 - 2u^3v$

b) $4u^2v - 4uv^2$

c) $2u^2v - 2uv^2$

d) $4uv^3 - 4u^3v$

_____ 2. Find the iterated integral for $\iint\limits_R dA$, where R is the first quadrant of the ellipse $4x^2 + 9y^2 = 36$ and the transformation is $x = 3u \cos v$, $y = 2u$ si v.

a) $\int_0^1 \int_0^{\pi/2} du\ dv$

b) $\int_0^1 \int_0^{\pi/2} 6u\ du\ dv$

c) $\int_0^1 \int_0^{\pi/2} 3u^2\ du\ dv$

d) $\int_0^1 \int_0^{\pi/2} u\ du\ dv$

_____ 1. For $\mathbf{F}(x, y) = (x + y)\mathbf{i} + (2x - 4y)\mathbf{j}$
which vector, **a**, **b**, **c**, or **d**, best
represents $\mathbf{F}(1, 1)$?

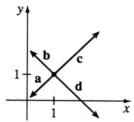

    a) **a**
    b) **b**
    c) **c**
    d) **d**

_____ 2. The gradient vector field for $f(x, y) = x^2 y^3$ is:

    a) $x^2\mathbf{i} + y^3\mathbf{j}$ b) $2x\mathbf{i} + 3y^2\mathbf{j}$

    c) $2xy^3\mathbf{i} + 3x^2 y^2\mathbf{j}$ d) $y^3\mathbf{i} + x^2\mathbf{j}$

_____ 3. If $\nabla f = \mathbf{F}$ the vector field $\mathbf{F}$ is:

    a) linear   b) conservative
    c) gradient   d) tangent

_____ **1.** A definite integral for $\int_C (x+y)ds$, where C is the curve given by $x = 3t$, $y = t$, $t \in [0, 1]$ is:

     a) $\int_0^1 4t \; dt$ b) $\int_0^1 4\sqrt{10} \; t \; dt$

     c) $\int_0^1 4t\sqrt{10t} \; dt$ d) $\int_0^1 4t\sqrt{t} \; dt$

_____ **2.** True, False:

The curve below appears to be piecewise smooth.

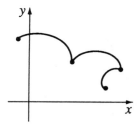

_____ **3.** A definite integral for $\int_C \mathbf{F} \; d\mathbf{r}$ where $\mathbf{F}(x, y) = x^2\mathbf{i} + y^2\mathbf{j}$ and C is the line from $(1, 0)$ to $(0, 1)$ given by $x = 1 - t$, $y = t$, $t \in [0, 1]$ is:

     a) $\int_0^1 (2t - 1)dt$ b) $\int_0^1 t^2 + (1 - t)^2 \; dt$

     c) $\int_0^1 t^3 + (1 - t)^2 t \; dt$ d) $\int_0^1 2 \; dt$

_____ **4.** True, False:

$\int_C f(x, y)dy = \int f(x(t), y(t))dt$.

_____ 1. For $f(x, y) = x^2 - y^2$ and C, the parabola $y = x^2$ from $(0, 0)$ to $(2, 4)$, $\int_C \nabla f \cdot d\mathbf{r} =$

     a) 0   b) 2   c) -2 d) -12

_____ 2. Which curve is simple but not closed?

     a)   b)

     c)   d)

_____ 3. True, False:
$\mathbf{F}(x, y) = (x^2 + 2y^2)\mathbf{i} + (y^2 + 2x^2)\mathbf{j}$ is conservative.

_____ 4. Let C be the line from $(1, 0)$ to $(0, 1)$ and
$\mathbf{F}(x, y) = y\mathbf{i} + (x + y)\mathbf{j}$. $\int_C \mathbf{F} \cdot d\mathbf{r} =$

     a) $\frac{1}{2}$ b) 1   c) 2   d) 4

_____ **1.** True, False:
If $\nabla f = \mathbf{F} = P\mathbf{i} + Q\mathbf{j}$, then
$\oint_C \mathbf{F} \cdot d\mathbf{r} = \iint\limits_D f(x, y)\, dA.$

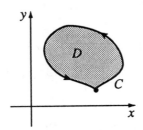

_____ **2.** Let C be the curve that bounds
the rectangle at the right.
For $\mathbf{F}(x, y) = (x^3 + y)\mathbf{i} + (2xy)\mathbf{j}$,
write a double integral for $\oint_C \mathbf{F} \cdot dr$.

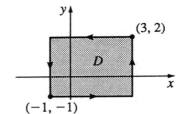

a) $\iint\limits_D (2y - 1)\, dA$

b) $\iint\limits_D (2xy - x^2 y)\, dA$

c) $\iint\limits_D (2x - 3x^2)\, dA$

d) $\iint\limits_D (x^3 + y - 2xy)\, dA$

_____ **3.** The area of a region D enclosed by curve C, such as that pictured in
question 1 above, may be described as:

a) $\frac{1}{2} \oint_C x \, dy$ 　　　　　 b) $\frac{1}{2} \oint_C y \, dx$

c) $\frac{1}{2} \oint_C x \, dy + y \, dx$ 　　　 d) $\frac{1}{2} \oint_C x \, dy - y \, dx$

_____ 1. Find the curl of $\mathbf{F}(x, y, z) = (x^2 + y^2)\mathbf{i} + (xz)\mathbf{j} + (yz)\mathbf{k}$.

   a) $-z\mathbf{i} + 2y\mathbf{j} + x\mathbf{k}$ b) $(z - x)\mathbf{i} + (z - 2y)\mathbf{k}$

   c) $(x - z)\mathbf{i} + 2x\mathbf{j} + 2y\mathbf{k}$ d) $y\mathbf{k}$

_____ 2. True, False:
   If $\mathbf{F}$ is conservative then curl $\mathbf{F} = \mathbf{0}$.

_____ 3. Find div $\mathbf{F}$ for $\mathbf{F}(x, y, z) = (x^2 + y^2)\mathbf{i} + (xz)\mathbf{j} + (yz)\mathbf{k}$.

   a) $2z - x - y$ b) $2x + 3y + 2z$

   c) $2x + y$ d) $y + x$

_____ 4. True, False: div curl $\mathbf{F} = 0$.

_____ 5. In vector form Green's Theorem says $\oint_C \mathbf{F} \cdot d\mathbf{r} =$

   a) $\iint\limits_D \mathrm{curl}\ \mathbf{F}\ dA$ b) $\iint\limits_D \mathrm{div}\ \mathbf{F}\ dA$

   c) $\iint\limits_D (\mathrm{curl}\ \mathbf{F}) \cdot \mathbf{k}\ dA$ d) $\iint\limits_D (\mathrm{div}\ \mathbf{F}) \cdot \mathbf{k}\ dA$

_____ 1. A parametrization of the cylinder at the right is:

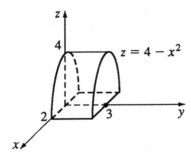

$z = 4 - x^2$

a) $x = u,\ y = v,\ z = v - u^2,\ -2 \le u \le 2,\ 0 \le v \le 3$
b) $x = u,\ y = 3,\ z = 4 - v^2,\ -2 \le u \le 2,\ 0 \le v \le z$
c) $x = u,\ y = v,\ z = 4 - u^2,\ -2 \le u \le 2,\ 0 \le v \le 3$
d) $x = u,\ z = 4 - v^2,\ -2 \le u \le 2,\ 0 \le v \le 2$

_____ 2. Find the normal vector to the tangent plane at the point corresponding to $(u,\ v) = (1,\ 2)$ on the surface given by $x = 2u,\ y = u^2 + v^2,\ z = 3v$.

a) $6\mathbf{i} + 6\mathbf{j} - 8\mathbf{k}$ b) $-6\mathbf{i} - 6\mathbf{j} + 8\mathbf{k}$
c) $6\mathbf{i} - 6\mathbf{j} + 8\mathbf{k}$ d) $-6\mathbf{i} + 6\mathbf{j} - 8\mathbf{k}$

_____ 3. A surface S is parameterized by $x = \sin u,\ y = \sin v,\ z = \cos v$ for $(u,\ v) \in D, 0 \le u \le \frac{\pi}{2},\ 0 \le v \le \frac{\pi}{2}$. A double integral for the surface area of S is:

a) $\iint\limits_{D} \sqrt{(\cos u \cos v)^2 + (\sin u \sin v)^2}\ dA$

b) $\iint\limits_{D} \sqrt{(\cos u \sin v)^2 + (\cos v \sin u)^2}\ dA$

c) $\iint\limits_{D} \cos u\ dA$

d) $\iint\limits_{D} \sqrt{3}\ dA$

_____ 1. Find an iterated integral for the surface integral of $f(x, y, z) = x + y$ over the surface S given by

$\mathbf{r}(u, v) = (u + v)\mathbf{i} + (u - v)\mathbf{j} + (uv)\mathbf{k}, \ 0 \le u \le 1, 0 \le v \le 1.$

a) $\int_0^1 \int_0^1 2u\sqrt{2u^2 + 2v^2} \ du \ dv$

b) $\int_0^1 \int_0^1 \sqrt{2u^2 + 2v^2} \ du \ dv$

c) $\int_0^1 \int_0^1 2u\sqrt{2u^2 + 2v^2 + 4} \ du \ dv$

d) $\int_0^1 \int_0^1 \sqrt{2u^2 + 2v^2 + 4} \ du \ dv$

_____ 2. Find an iterated integral for $\iint\limits_S \mathbf{F} \cdot d\mathbf{S}$ where S is the surface

$\mathbf{r}(u, v) = u^2\mathbf{i} + v^2\mathbf{j} + uv\mathbf{k}, \ 0 \le u \le 1, 0 \le v \le 1$, and
$\mathbf{F}(x, y, z) = x\mathbf{i} + y\mathbf{j} + z\mathbf{k}.$

a) $\int_0^1 \int_0^1 2uv \ du \ dv$

b) $\int_0^1 \int_0^1 4u^2v^2 \ du \ dv$

c) $\int_0^1 \int_0^1 (2u^2 + 2v^2) \ du \ dv$

d) $\int_0^1 \int_0^1 (2u^2 + 2v^2 + 4u^2v^2) \ du \ dv$

_____ 1. The surface S has as its boundary the simple closed curve C. By Stokes' Theorem $\iint_S \text{curl } \mathbf{F} \cdot d\mathbf{S} =$

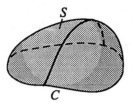

a) $\int_C \mathbf{F} \cdot d\mathbf{S}$

b) $\int_C \mathbf{F} \cdot d\mathbf{r}$

c) $\int_C \text{div } \mathbf{F} \cdot d\mathbf{S}$

d) $\int_C \text{div } \mathbf{F} \cdot d\mathbf{r}$

_____ 2. Use Stokes' Theorem to rewrite $\int_C \mathbf{F} \cdot d\mathbf{r}$, where $F(x, y, z) = z\mathbf{i} - x\mathbf{j} + y\mathbf{k}$ and C is the circle $x^2 + y^2 = 36$, $z = 0$. Use the surface S: $z = 36 - x^2 - y^2$, $z \geq 0$.

a) $\int_{-6}^{6} \int_{-\sqrt{36-u^2}}^{\sqrt{36-u^2}} (2u + 2v - 1) dv\, du$

b) $\int_{-6}^{6} \int_{-\sqrt{36-u^2}}^{\sqrt{36-u^2}} (2u - 2v - 1) dv\, du$

c) $\int_{-6}^{6} \int_{-\sqrt{36-u^2}}^{\sqrt{36-u^2}} (2u + 2v + 1) dv\, du$

d) $\int_{-6}^{6} \int_{-\sqrt{36-u^2}}^{\sqrt{36-u^2}} (2u - 2v + 1) dv\, du$

_____ **1.** Let S be the surface of the closed cylindrical "half can" below and
$\mathbf{F}(x,\ y,\ z) = 2xy^2\mathbf{i} + xz^2\mathbf{j} + yz\mathbf{k}$. Write a triple iterated integral for
$\iint\limits_S \mathbf{F} \cdot d\mathbf{S}$.

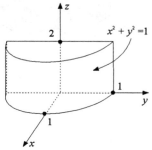

a) $\int_0^2 \int_0^1 \int_{-\sqrt{1-x^2}}^{\sqrt{1-x^2}} (4xy + 2xz + y)dy\ dx\ dz$

b) $\int_0^2 \int_0^1 \int_{-\sqrt{1-x^2}}^{\sqrt{1-x^2}} (2y^2 - z)dy\ dx\ dz$

c) $\int_0^2 \int_{-1}^1 \int_0^{\sqrt{1-x^2}} (4xy + 2xz + y)dx\ dy\ dz$

d) $\int_0^2 \int_{-1}^1 \int_0^{\sqrt{1-y^2}} (2y^2 + y)dx\ dy\ dz$

Section 15.1

_____ **1.** Is the figure at the right the graph of a direction field?

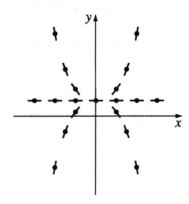

_____ **2.** The general solution to $y' = \frac{y}{x}$, for x, $y > 0$ is:

a) $y = Cx$ b) $y = x + C$ c) $y' = \ln x + C$ d) $y' = e^{x+C}$

_____ **3.** Solve $y\frac{dy}{dx} = -3x^2$, $(y > 0)$ with initial condition $y(0) = 2$.

a) $y = 4e^{-x^3}$ b) $y = 4 - 2x^3$ c) $y = \sqrt{4 - 2x^3}$ d) $y = 4e^{-2x^3}$

_____ **4.** By Euler's method the first approximated value for $\frac{dy}{dx} = x - 4y$, $y(11) = 2$ with step size 0.5 is:

a) $(11.0, 3.5)$ b) $(11.5, 3.5)$ c) $(11.0, 3.75)$ d) $(11.5, 3.75)$

_____ **5.** True, False: $y' = \frac{4xy^2 + 6xy + y^3}{x^3 + x^2y}$ is homogeneous.

Section 15.2

_____ **1.** True, False: $xy' + 6x^2y = 10 - x^3$ is linear.

_____ **2.** What is the integrating factor for $xy' + 6x^2y = 10 - x^3$?

a) e^{3x^2} b) e^{6x} c) e^{3x^3} d) e^{6x^2}

_____ **3.** The solution to $x\frac{dy}{dx} - 2y = x^3$ is:

a) $y = e^{x^3} + Cx^2$ b) $y = e^{x^3} + x^2 + C$

c) $y = x^3 + x^2 + C$ d) $y = x^3 + Cx^2$

Section 15.3

_____ 1. True, False:
$xe^{xy^2} + y + (e^{xy^2} + x)y' = 0$ is exact.

_____ 2. The solution to $\frac{dy}{dx} = \frac{\cos x \cos y + y}{\sin x \sin y - x}$ is given implicitly by:

   a) $\sin x \sin y + \cos x \cos y + xy = C$
   b) $\sin x \cos y + xy = C$
   c) $\sin y \cos x + xy = C$
   d) $\sin x \cos y - \cos x \sin y + xy = C$

_____ 3. An integrating factor for $(6y^2 + 3xy^4) + (6xy + 4x^2y^3)y' = 0$ is:

   a) $I(x, y) = x$ b) $I(x, y) = x^2$

   c) $I(x, y) = y$ d) $I(x, y) = y^2$

_____ 4. True, False:
If $\frac{dy}{dx} = F(x, y)$ is separable, then it is exact.

Section 15.4

_____ **1.** True, False:
$xy' + x^2 y^2 = 1$ is separable.

_____ **2.** True, False:
$y' = \frac{2xy - y^2 + y}{x^2 - 2xy + x}$ is exact.

_____ **3.** True, False:
$x^2 y + xy' = 10 - x$ is linear.

_____ **4.** True, False:
$y' = \frac{6x^2 y^2 + xy^3 + y^4}{x^3 + x^2 y^2}$ is homogeneous.

Section 15.5

_____ 1. True, False:
If y_1 and y_2 are linearly independent solutions to
$P(x)y'' + Q(x)y' + R(x)y = 0$, then all solutions have the form
$c_1 y_1 + c_2 y_2$.

_____ 2. The general solution to $y'' - 6y' + 8y = 0$ is:

a) $y = e^x(c_1 \cos 4x + c_2 \sin 4x)$
b) $y = e^{4x}(c_1 \cos x + c_2 \sin x)$
c) $y = c_1 e^{2x} + c_2 e^{4x}$
d) $y = c_1 e^{2x} + c_2 x e^{4x}$

_____ 3. The solution to $y'' - 10y' + 25y = 0$, $y(0) = 5$, $y(1) = 0$ is:

a) $y = 5e^{5x} - xe^{5x}$ b) $y = 5e^{5x} - 5xe^{5x}$

c) $y = e^{5x} - 5xe^{5x}$ d) $y = 5e^{5x} + xe^{5x}$

_____ 4. True, False:
An initial-value problem specifies $y(x_0) = y_0$ and $y(x_1) = y_1$ in the
statement of the problem.

_____ 1. If $y_c(x)$ is the general solution to $ay'' + by' + cy = 0$ and $y_p(x)$ is a particular solution to $ay'' + by' + cy = G(x)$, then the general solution to the nonhomogeneous equation is $y(x) =$

    a) $y_c(x) + y_p(x)$ b) $y_c(x) - y_p(x)$

    c) $y_p(x) - y_c(x)$ d) $y_c(x)y_p(x)$

_____ 2. By the method of undetermined coefficients, a particular solution to $y'' - 5y' + 6y = e^{4x}$ will have the form:

    a) $y_p(x) = Ae^x$ b) $y_p(x) = A_1e^x + A_2e^{6x}$

    c) $y_p(x) = A_1e^{2x} + A_2e^{3x}$ d) $y_p(x) = Ae^{4x}$

_____ 3. By the method of variation of parameters, a particular solution to $y'' - 2y' + y = 3e^{4x}$ is $u_1y_1 + u_2y_2$, where $y_1 = e^x$, $y_2 = xe^x$, and

    a) $u_1 = -xe^{3x} + e^{3x}$, $u_2 = -e^{3x}$

    b) $u_1 = -xe^{3x}$, $u_2 = e^{3x}$

    c) $u_1 = -xe^{3x}$, $u_2 = xe^{3x} - e^{3x}$

    d) $u_1 = -xe^{3x} + \frac{1}{3}e^{3x}$, $u_2 = e^{3x}$

Section 15.7

_____ **1.** A series circuit has a resistor with $R = 15\Omega$, an inductor with $L = 2$, a capacitor with $C = 0.005$ F and a 6 volt battery. An equation for the charge Q at time t is:

a) $2Q'' + 200Q' + 15Q = 6$
b) $2Q'' + 6Q' + 15Q = 200$
c) $2Q'' + 15Q' + 200Q = 6$
d) $2Q'' + 15Q' + 6Q = 200$

_____ **2.** A spring with a 10-kg mass is immersed in a fluid with damping constant 60. To maintain the spring stretched 3 m beyond its equilibrium requires a force of 150 N. The general solution to the equation modeling the motion is:

a) $x = c_1 e^t + c_2 e^{5t}$
b) $x = c_1 e^{-t} + c_2 e^{-5t}$
c) $x = e^{-t}(c_1 \cos 5t + c_2 \sin 5t)$
d) $x = e^{-5t}(c_1 \cos t + c_2 \sin t)$

_____ 1. Find the recursion relation among the coefficients of $y = \sum\limits_{n=0}^{\infty} a_n x^n$ that is the series solution to $y'' = 2y$.

    a) $a_n = \dfrac{2a_{n-2}}{n(n-1)}$

    b) $a_n = \dfrac{2^n a_{n-2}}{n(n-1)}$

    c) $a_n = \dfrac{2a_{n-1}}{n-1}$

    d) $a_n = \dfrac{2^n a_{n-1}}{n-1}$

_____ 2. The series solution to $y' = xy$ is:

    a) $y = \sum\limits_{n=0}^{\infty} \dfrac{a_0}{2^n n!} x^{2n}$

    b) $y = \sum\limits_{n=0}^{\infty} \dfrac{a_0}{n!} x^{2n}$

    c) $y = \sum\limits_{n=0}^{\infty} \dfrac{2^n a_0}{n!} x^{2n}$

    d) $y = \sum\limits_{n=0}^{\infty} \dfrac{n! a_0}{2^n} x^{2n}$

Answers to On Your Own

Section 10.1
1. B
2. Sometimes
3. Sometimes
4. Always
5. False
6. A
7. D

Section 10.2
1. False
2. True
3. B
4. False
5. C
6. True

Section 10.3
1. C
2. False
3. No
4. Yes
5. B

Section 10.4
1. Never
2. Yes
3. No
4. Yes
5. A

Section 10.5
1. Yes
2. Yes
3. D
4. False

Section 10.6
1. False
2. True
3. True
4. A
5. B

Section 10.7
1. True
2. False
3. True
4. True

Section 10.8
1. Sometimes
2. True
3. C
4. B
5. C

Section 10.9
1. C
2. B
3. C

Section 10.10
1. C
2. C
3. A
4. True
5. B

Section 10.11
1. B
2. A
3. D

Section 10.12
1. B
2. A

Section 11.1
1. A
2. C
3. D
4. D
5. D

Section 11.2
1. B
2. C
3. D
4. False
5. C

Section 11.3
1. B
2. A
3. B
4. B

Section 11.4
1. D
2. False
3. B
4. D
5. C

Section 11.5
1. D
2. True
3. A
4. D
5. C

Section 11.6
1. B
2. D
3. C
4. B
5. B

Answers to On Your Own

Section 11.7	Section 11.8	Section 11.9	Section 11.10
1. A	1. A	1. B	1. A
2. B	2. D	2. C	2. D
3. A	3. A	3. C	3. D
4. False	4. True	4. A	4. D
5. C	5. A		5. A

Section 12.1	Section 12.2	Section 12.3	Section 12.4
1. C	1. True	1. B	1. False
2. B	2. A	2. B	2. A
3. B	3. D	3. A	3. C
4. C	4. D	4. Sometimes	4. C
	5. False	5. D	

Section 12.5	Section 12.6	Section 12.7	Section 12.8
1. D	1. B	1. A	1. A
2. A	2. True	2. B	2. A
3. A	3. B	3. B	
4. C	4. C	4. A	
	5. D	5. False	

Section 13.1	Section 13.2	Section 13.3
1. A	1. B	1. D
2. C	2. False	2. False
3. Sometimes	3. B	3. B
		4. B

Section 13.4	Section 13.5	Section 13.6
1. B	1. A	1. C
2. D	2. D	

Section 13.7	Section 13.8	Section 13.9
1. False	1. A	1. D
2. A	2. B	2. D
3. A		
4. B		

Section 14.1	Section 14.2	Section 14.3
1. D	1. B	1. D
2. D	2. True	2. C
3. B	3. A	3. False
	4. False	4. A

Answers to On Your Own

Section 14.4
1. False
2. A
3. D

Section 14.5
1. B
2. True
3. C
4. True
5. C

Section 14.6
1. C
2. C
3. C

Section 14.7
1. C
2. B

Section 14.8
1. B
2. A

Section 14.9
1. D

Section 15.1
1. No
2. A
3. C
4. B
5. False

Section 15.2
1. True
2. A
3. D

Section 15.3
1. False
2. B
3. A
4. True

Section 15.4
1. False
2. False
3. True
4. False

Section 15.5
1. True
2. C
3. B
4. False

Section 15.6
1. A
2. D
3. D

Section 15.7
1. C
2. B

Section 15.8
1. A
2. A